AURORÆ

LECTOR HOUSE PUBLIC DOMAIN WORKS

AURORÆ

J. RAND CAPRON

ISBN: 978-93-89701-73-9

Published: 1879

LECTOR HOUSE LLP
E-MAIL: lectorpublishing@gmail.com

AURORÆ:
THEIR CHARACTERS AND SPECTRA.

BY

J. RAND CAPRON, F.R.A.S.

"And now the Northern Lights begin to burn, faintly at first, like sunbeams playing in the waters of the blue sea. Then a soft crimson glow tinges the heavens. There is a blush on the cheek of night. The colours come and go; and change from crimson to gold, from gold to crimson. The snow is stained with rosy light. Twofold from the zenith, east and west, flames a fiery sword; and a broad band passes athwart the heavens, like a summer sunset. Soft purple clouds come sailing over the sky, and through their vapoury folds the winking stars shine white as silver. With such pomp as this is Merry Christmas ushered in, though only a single star heralded the first Christmas." —LONGFELLOW.

1879.

TO

PROF. CHARLES PIAZZI SMYTH, F.R.S.E.,

ASTRONOMER ROYAL FOR SCOTLAND,

ONE OF THE EARLIEST SPECTROSCOPIC OBSERVERS

OF

THE AURORA AND ZODIACAL LIGHT,

THIS VOLUME

IS RESPECTFULLY DEDICATED

BY THE AUTHOR.

PREFACE.

Probably few of the phenomena of Nature so entirely charm and interest scientific and non-scientific observers alike as the Aurora Borealis, or "Northern Lights" as it is popularly called. Whether contemplated as the long low quiescent arc of silver light illuminating the landscape with a tender radiance, as broken clouds and columns of glowing ruddy light, or as sheaves of golden rays, aptly compared by old writers to aerial spears, such a spectacle cannot fail at all times to be a subject of admiration, in some cases even of awe.

Hence it is no wonder that the Aurora has always received a considerable amount of attention at the hands of scientific men. Early explorers of the Arctic Regions made constant and important observations of it and its character; and the list of references to works given in the Appendix will show how often it formed the subject of monographs and communications to learned Societies. The early contributions seem relatively more numerous than those of a later date; and the substance of them will be found well summed up in Dr. Brewster's 'Edinburgh Encyclopædia' (1830), article "Aurora." A most complete and able epitome of our more recent experience and knowledge of the Aurora and its spectrum has been contributed by my friend Mr. Henry R. Procter to the present (9th) edition of the 'Encyclopædia Britannica,' article "Aurora Polaris." It is, however, a drawback to Encyclopædic articles that their matter is of necessity condensed, and that they rarely have the very desirable aid of drawings and engravings to illustrate their subjects. In spite, therefore, of the exhaustive way, both as to fact and theory, in which the contributor to the 'Encyclopædia Britannica' has realized his task, it seemed to me there was still room left for a popular treatise, having for its object the description of Auroræ, their characters and spectra. The question of the Aurora spectrum seems the more worthy of extended discussion in that it still remains an unsolved problem. In spite of the observations and researches of Ångström, Lemström, and Vogel abroad, and of Piazzi Smyth, Herschel, Procter, Backhouse, and others at home, the goal is not yet reached; for while the faint and more refrangible lines are but doubtfully referred to air, the bright and sharp red and green lines, which mainly characterize the spectrum, are as yet unassociated with any known analogue.

With these views, and to incite to further and closer observations, I have been induced to publish the present volume as a sort of Auroral Guide. For much of the history of the Aurora I am indebted to, and quote from former articles and records, including the two excellent Encyclopædic ones before referred to. Mr. Procter, Mr. Backhouse, and my friend Mr. W. H. Olley have each kindly furnished me with much in the way of information and suggestion. Dr. Schuster has lent me tubes showing the true oxygen spectrum; while Herr Carl Bock, the Norwegian naturalist, has enabled me to reproduce a veritable curiosity, viz. a picture in oil painted by the light of a Lapland Aurora. The experiments detailed in Part III. were suggested by the earlier ones of De la Rive, Varley, and others, and demonstrate the effect of the magnet on electric discharges. For assistance in these I am indebted to my friend Mr. E. Dowlen.

The illustrations are mainly from original drawings of my own. Those from other sources are acknowledged. Messrs. Mintern have well reproduced in chromo-lithography the coloured drawings illustrating the Auroræ, moon-patches, &c.

CONTENTS

PART III.
MAGNETO-ELECTRIC EXPERIMENTS IN CONNEXION WITH THE AURORA.

LIST OF PLATES

PART I.
THE AURORA AND ITS CHARACTERS.

CHAPTER I.

THE AURORA AS KNOWN TO THE ANCIENTS.

Seneca's 'Quæstiones Naturales,' Lib. I. c. xiv. Description of Auroræ.

In Seneca's 'Quæstiones Naturales,' Lib. I. c. xiv., we find the following:—"Tempus est, alios quoque ignes percurrere, quorum diversæ figuræ sunt. Aliquando emicat stella, aliquando ardores sunt, aliquando fixi et hærentes, nonnunquam volubiles. Horum plura genera conspiciantur. Sunt *Bothynoë*[1], quum velut corona cingente introrsus igneus cœli recessus est similis effossæ in orbem speluncæ. Sunt *Pithitæ*[2], quum magnitudo vasti rotundique ignis dolio similis, vel fertur vel in uno loco flagrat. Sunt *Chasmata*[3], quum aliquod cœli spatium desedit, et flammam dehiscens, velut in abdito, ostentat. Colores quoque omnium horum plurimi sunt. Quidam ruboris acerrimi, quidam evanidæ ac levis flammæ, quidam candidæ lucis, quidam micantes, quidam æqualiter et sine eruptionibus aut radiis fulvi.

* * *

Seneca, c. xv.

C. xv. "Inter hæc ponas licet et quod frequenter in historiis legimus, cœlum ardere visum: cujus nonnunquam tam sublimis ardor est ut inter ipsa sidera videatur, nonnunquam tam humilis ut speciem longinqui incendii præbeat.

"Sub Tiberio Cæsare cohortes in auxilium Ostiensis coloniæ cucurrerunt, tanquam conflagrantis, quum cœli ardor fuisset per magnam partem noctis, parum lucidus crassi fumidique ignis."

Translation.

We may translate this:—"It is time other fires also to describe, of which there are diverse forms.

"Sometimes a star shines forth; at times there are fire-glows, sometimes fixed and persistent, sometimes flitting. Of these many sorts may be distinguished. There are Bothynoë, when, as within a surrounding corona, the fiery recess of the sky is like to a cave dug out of space. There are Pithitæ, when the expanse of a vast and rounded fire similar to a tub (dolium) is either carried about or glows in one

---

[1] βόθυνος, a hollow.

[2] πίθος, a cask.

[3] χάσμς, a chasm.

spot.

"There are Chasmata, when a certain portion of the sky opens, and gaping displays the flame as in a porch. The colours also of all these are many. Certain are of the brightest red, some of a flitting and light flame-colour, some of a white light, others shining, some steadily and yellow without eruptions or rays.

* * *

"Amongst these we may notice, what we frequently read of in history, the sky is seen to burn, the glow of which is occasionally so high that it may be seen amongst the stars themselves, sometimes so near the Earth (humilis) that it assumes the form of a distant fire. Under Tiberius Cæsar the cohorts ran together in aid of the colony of Ostia as if it were in flames, when the glowing of the sky lasted through a great part of the night, shining dimly like a vast and smoking fire."

Auroræ frequently read of in history.

From the above passages many striking particulars of the Aurora may be gathered; and by the division of the forms of Aurora into classes it is evident they were, at that period, the subject of frequent observation. The expression "et quod frequenter in historiis legimus" shows, too, that the phenomena of Auroral displays were a matter of record and discussion with the writers of the day.

Various forms of Aurora may be recognized in the passages from Chap. xiv.; while in those from Chap. xv. a careful distinction is drawn between the Auroræ seen in the zenith or the upper regions of the sky, and those seen on the horizon or apparently (and no doubt in some cases actually) near the Earth's surface.

A spurious Aurora.

The description of the cohorts running to the fire only to find it an Aurora, calls to mind the many similar events happening in our own days. Not, however, but that a mistake may sometimes occur in an opposite direction. In the memoirs of Baron Stockmar an amusing anecdote is related of one Herr von Radowitz, who was given to making the most of easily picked up information. A friend of the Baron's went to an evening party near Frankfort, where he expected to meet Herr von Radowitz. On his way he saw a barn burning, stopped his carriage, assisted the people, and waited till the flames were nearly extinguished. When he arrived at his friend's house he found Herr von Radowitz, who had previously taken the party to the top of the building to see an Aurora, dilating on terrestrial magnetism, electricity, and so forth. Radowitz asked Stockmar's friend, "Have you seen the beautiful Aurora Borealis?" He replied, "Certainly; I was there myself; it will soon be over." An explanation followed as to the barn on fire: Radowitz was silent some ten minutes, then took up his hat and quietly disappeared.

Auroræ as portents.

It is probable that many of the phantom combats which are recorded to have appeared in forms of fire in the air on the evenings preceding great battles might be traced to Auroræ, invested with distinct characteristics by the imagination of the beholders. Auroræ are said to have appeared in the shape of armies of horse and foot engaged in battle in the sky before the death of Julius Cæsar, which they were supposed to foretell. For more than a year before the siege and destruction of Jerusalem by Titus Vespasian, the Aurora was said to have been frequently visible in Palestine.

Josephus, in his 'Wars of the Jews' (Whiston's Translation, Book VI. chap. v. sect. 3), in referring to the signs and wonders preceding the destruction of Jerusalem, speaks of a star or comet, and that a great light shone round about the altar and the holy house, which light lasted for half an hour, and that a few days after the feast of unleavened bread a certain prodigious and incredible phenomenon appeared—"for before sunsetting chariots and troops of soldiers in their armour were seen running about among the clouds, and surrounding of cities." (This, if an Aurora, must have been an instance of a daylight one.)

We find in Book II. of Maccabees, chap. v. verses 1, 2, 3, 4 (B.C. about 176 years):—

"1. About this same time Antiochus prepared his second voyage into Egypt:

"2. And then it happened that through all the city, for the space almost of forty days, there were seen horsemen running in the air, in cloth of gold, and armed with lances like a band of soldiers.

"3. And troops of horsemen in array, encountering and running one against another, with shaking of shields and multitude of pikes, and drawing of swords and casting of darts, and glittering of golden ornaments and harness of all sorts.

"4. Wherefore every man prayed that that apparition might turn to good."

Early descriptions of Auroræ.

In Aristotle's 'De Meteoris,' Lib. I. c. iv. and v., the Aurora is described as an appearance resembling flame mingled with smoke, and of a purple red or blue colour. Pliny (Lib. II. c. xxvii.) speaks of a bloody appearance of the heavens which seemed like a fire descending on the earth, seen in the third year of the 107th Olympiad, and of a light seen in the nighttime equal to the brightness of the day, in the Consulship of Cæcilius and Papirius (Lib. II. c. xxxiii.), both of which may be referred to Auroræ.

In the 'Annals of Philosophy,' vol. ix. p. 250, it is stated that the Aurora among English writers is first described by Matthew of Westminster, who relates that in A.D. 555 lances were seen in the air ("quasi species lancearum in aëre visæ sunt a septentrionali usque ad occidentem").

In the article in the 'Edinb. Encyc.' vol. iii. (1830), the Aurora (known to the vulgar as "streamers" or "merry dancers") is distinguished in two kinds—the "tranquil" and the "varying." Musschenbroek enumerates as forms:—*trabs*, "the beam," an oblong tract parallel to the horizon; *sagitta*, "the arrow;" *faces*, "the torch;" *capra saltans*, "the dancing goat;" *bothynoë*, "the cave," a luminous cloud having the appearance of a recess or hollow in the heavens, surrounded by a corona; *pithiæ*, "the tun," an Aurora resembling a large luminous *cask*. The two sorts of Auroræ distinguished as the "bothynoë" and "pithiæ" are evidently taken from the passage in Seneca's 'Quæstiones' before quoted. In 'Liberti Fromondi Meteorologicorum' (London, 1656), Lib. II. cap. v. "De Meteoris supremæ regionis aëris," art. 1. De Capra, Trabe, Pyramide, &c., these and other fantastic forms attributable to Auroræ are more fully described.

In the article "Aurora Polaris," Encyc. Brit. edit. ix., we find noted that from a curious passage in Sirr's 'Ceylon and the Cingalese,' vol. ii. p. 117, it would seem that the Aurora, or something like it, is visible occasionally in Ceylon, where the natives call it "Buddha Lights," and that in many parts of Ireland a scarlet Aurora is supposed to be a shower of blood. The earliest mentioned Aurora (in Ireland) was in 688, in the 'Annals of Cloon-mac-noise,' after a battle between Leinster and Munster, in which Foylcher O'Moyloyer was slain.

In the article in the Edinb. Encyc. before referred to it is stated that it was not much more than a century ago that the phenomenon had been noticed to occur with frequency in our latitudes.

Dr. Halley had begun to despair of seeing one till the fine display of 1716.

Early notices of Auroræ not frequent in our latitudes.

The first account on record in an English work is said to be in a book entitled 'A Description of Meteors by W. F. D. D.' (reprinted, London, 1654), which speaks of "burning spears" being seen January 30, 1560. The next is recorded by Stow as occurring on October 7, 1564; and, according to Stow and Camden, an Aurora was seen on two nights, 14th and 15th November, 1574.

Twice, again, an Aurora was seen in Brabant, 13th February and 28th September, 1575. Cornelius Gemma compared these to spears, fortified cities, and armies fighting in the air. Auroræ were seen in 1580 and 1581 in Wirtemberg, Germany.

Then we have no record till 1621, when an Aurora, described by Gassendi in his 'Physics,' was seen all over France, September 2nd of that year.

In November 1623 another, described by Kepler, was seen all over Germany.

From 1666 to 1716 no appearance is recorded in the 'Transactions of the French Academy of Sciences;' but in 1707 one was seen in Ireland and at Copenhagen; while in 1707 and 1708 the Aurora was seen five times.

The Aurora of 1716, occurring after an interval of eighty years, which Dr. Halley describes, was very brilliant and extended over much country, being seen from the west of Ireland to the confines of Russia and the east of Poland, extending nearly 30° of longitude, and from about the 50th degree of latitude, over almost all the north of Europe, and in all places exhibiting at the same time appearances similar to those observed in London. An Aurora observed in Bologna in 1723 was stated to be the first that had ever been seen there; and one recorded in the 'Berlin Miscellany' for 1797 is called a very unusual phenomenon. Nor did Auroræ appear more frequent in the Polar Regions at that time, for Cælius states that the oldest inhabitants of Upsala considered the phenomenon as quite rare before 1716. Anderson, of Hamburg, writing about the same time, says that in Iceland the inhabitants themselves were astonished at the frequent Auroræ then beginning to take place; while Torfæus, the Icelander, who wrote in 1706, was old enough to remember the time when the Aurora was an object of terror in his native country.

According to M. Mairan, 1441 Auroræ were observed between A.D. 583 and 1751, of which 972 were observed in the winter half-years and 469 during the summer half-years. In our next Chapter we propose to give some general descriptions of Auroræ from comparatively early sources.

CHAPTER II.

SOME GENERAL DESCRIPTIONS OF AURORÆ.

Sir John Franklin's description.

Sir John Franklin ('Narrative of a Journey to the Shores of the Polar Sea in the years 1819, 1820, 1821, 1822') describes an Aurora in these terms:—

Parts of the Aurora: beams, flashes, and arches.

"For the sake of perspicuity I shall describe the several parts of the Aurora, which I term beams, flashes, and arches.

"The beams are little conical pencils of light, ranged in parallel lines, with their pointed extremities towards the earth, generally in the direction of the dipping-needle.

Formation of the Aurora.

"The flashes seem to be scattered beams approaching nearer to the earth, because they are similarly shaped and infinitely larger. I have called them flashes, because their appearance is sudden and seldom continues long. When the Aurora first becomes visible it is formed like a rainbow, the light of which is faint, and the motion of the beams undistinguishable. It is then in the horizon. As it approaches the zenith it resolves itself into beams which, by a quick undulating motion, project themselves into wreaths, afterwards fading away, and again and again brightening without any visible expansion or contraction of matter. Numerous flashes attend in different parts of the sky."

Arches of the Aurora.

Sir John Franklin then points out that this mass would appear like an arch to a person situated at the horizon by the rules of perspective, assuming its parts to be equidistant from the earth; and mentions a case when an Aurora, which filled the sky at Cumberland House from the northern horizon to the zenith with wreaths and flashes, assumed the shape of arches at some distance to the southward. He then continues:—"But the Aurora does not always make its first appearance as an arch. It sometimes rises from a confused mass of light in the east or west, and crosses the sky towards the opposite point, exhibiting wreaths of beams or coronæ boreales on its way. An arch also, which is pale and uniform at the horizon, passes the zenith without displaying any irregularity or additional brilliancy." Sir

John Franklin then mentions seeing three arches together, very near the northern horizon, one of which exhibited beams and even colours, but the other two were faint and uniform. (See example of a doubled arc Aurora observed at Kyle Akin, Skye, Plate VII.)

He also mentions an arch visible to the southward exactly similar to one in the north. It appeared in fifteen minutes, and he suggests it probably had passed the zenith before sunset. The motion of the whole body of the Aurora from the northward to the southward was at angles not more than 20° from the magnetic meridian. The centres of the arches were as often in the magnetic as in the true meridian. A delicate electrometer, suspended 50 feet from the ground, was never perceptibly affected by the Aurora.

Aurora does not often appear until some hours after sunset.

Sir John Franklin further remarks that the Aurora did not often appear immediately after sunset, and that the absence of that luminary for some hours was in general required for the production of a state of atmosphere favourable to the generation of the Aurora.

Aurora seen in daylight.

On one occasion, however (March 8th, 1821), he observed it distinctly previous to the disappearance of daylight; and he subsequently states that on four occasions the coruscations of the Aurora were seen very distinctly before daylight had disappeared.

[In the article "Aurora Polaris," Encyc. Brit. edit. ix., the Transactions of the Royal Irish Academy, 1788, are referred to, where Dr. Usher notices that the Aurora makes the stars flutter in the telescope; and that, having remarked this effect strongly one day at 11 A.M., he examined the sky, and saw an Auroral corona with rays to the horizon.

Instances are by no means rare of the principal Aurora-line having been seen in waning sunlight, and in anticipation of an Aurora which afterwards appeared.]

The Rev. James Farquharson's observations. Auroral arch. Passage across the zenith.

The Rev. James Farquharson, from the observation of a number of Auroræ in Aberdeenshire in 1823 ('Philosophical Transactions,' 1829), concluded:—that the Aurora follows an invariable order in its appearance and progress; that the streamers appear first in the north, forming an arch from east to west, having its vertex at the line of the magnetic meridian (when this arch is of low elevation it is of considerable breadth from north to south, having the streamers placed crosswise in relation to its own line, and all directed towards a point a little south of the zenith); that the arch moves forward towards the south, contracting laterally as it approaches the zenith, and increasing its intensity of light by the shortening of the streamers and the gradual shifting of the angles which the streamers near the east and west extremities of the arch make with its own line, till at length these streamers become parallel to that line, and then the arch is seen in a narrow belt 3°

or 4° only in breadth, stretching across the zenith at right angles to the magnetic meridian; that it still makes progress southwards, and after it has reached several degrees south of the zenith again enlarges its breadth by exhibiting an order of appearances the reverse of that which attended its progress towards the zenith from the north; that the only conditions that can explain and reconcile these appearances are that the streamers of the Aurora are vertical, or nearly so, and form a deep fringe which stretches a great way from east to west at right angles to the magnetic meridian, but which is of no great thickness from north to south, and that the fringe moves southward, preserving its direction at right angles to the magnetic meridian.

M. Lottin's observations.

Dr. Lardner, in his 'Museum of Science and Art,' vol. x. p. 189 *et seq.*, alludes to a description of "this meteor" (*sic*) supplied by M. Lottin, an officer of the French Navy, and a Member of the Scientific Commission to the North Seas. Between September 1838 and April 1839, being the interval when the sun was constantly below the horizon, this savant observed nearly 150 Auroræ. During this period sixty-four were visible, besides many concealed by a clouded sky, but the presence of which was indicated by the disturbances they produced upon the magnetic needle.

The succession of appearances and changes presented by these "meteors" is thus graphically described by M. Lottin:—

Formation of the auroral bow.

"Between four and eight o'clock P.M. a light fog, rising to the altitude of six degrees, became coloured on its upper edge, being fringed with the light of the meteor rising behind it. This border, becoming gradually more regular, took the form of an arch, of a pale yellow colour, the edges of which were diffuse, the extremities resting on the horizon. This bow swelled slowly upwards, its vertex being constantly on the magnetic meridian. Blackish streaks divided regularly the luminous arc, and resolved it into a system of rays. These rays were alternately extended and contracted, sometimes slowly, sometimes instantaneously, sometimes they would dart out, increasing and diminishing suddenly in splendour. The inferior parts, or the feet of the rays, presented always the most vivid light, and formed an arc more or less regular. The length of these rays was very various, but they all converged to that point of the heavens indicated by the direction of the southern pole of the dipping-needle. Sometimes they were prolonged to the point where their directions intersected, and formed the summit of an enormous dome of light.

It ascends to the zenith. Reaches the zenith.

"The bow then would continue to ascend toward the zenith. It would suffer an undulatory motion in its light—that is to say, that from one extremity to the other the brightness of the rays would increase successively in intensity. This luminous current would appear several times in quick succession, and it would pass much more frequently from west to east than in the opposite direction. Sometimes, but rarely, a retrograde motion would take place immediately afterward; and as soon as this wave of light had run successively over all the rays of the Aurora from west to east, it would return in the contrary direction to the point of its departure, producing such an effect that it was impossible to say whether the rays themselves were actually affected by a motion of translation in a direction nearly horizontal, or if this more vivid light was transferred from ray to ray, the system of rays themselves suffering no change of position. The bow, thus presenting the appearance of an alternate motion in a direction nearly horizontal, had usually the appearance of the undulations or folds of a ribbon or flag agitated by the wind. Sometimes one, and sometimes both of its extremities would desert the horizon, and then its folds would become more numerous and marked, the bow would change its character and assume the form of a long sheet of rays returning into itself, and consisting of several parts forming graceful curves. The brightness of the rays would vary suddenly, sometimes surpassing in splendour stars of the first magnitude; these rays would rapidly dart out, and curves would be formed and developed like the folds of a serpent; then the rays would affect various colours, the base would be red, the middle green, and the remainder would preserve its clear yellow hue. Such was the arrangement which the colours always preserved. They were of admirable transparency, the base exhibiting blood-red, and the green of the middle being that of the pale emerald; the brightness would diminish, the colours disappear and all be extinguished, sometimes suddenly and sometimes by slow degrees. After this disappearance fragments of the bow would be reproduced, would continue their upward movement and approach the zenith; the rays, by the effect of perspective, would be gradually shortened; the thickness of the arc, which presented then the appearance of a large zone of parallel rays, would be extended; then the vertex of the bow would reach the magnetic zenith, or the point to which the south pole of the dipping-needle is directed. At that moment the rays would be seen in the direction of their feet. If they were coloured they would appear as a large red band, through which the green tints of their superior parts could be distinguished, and if the wave of light above mentioned passed along them their feet would form a long sinuous undulating zone; while throughout all these changes the rays would never suffer any oscillation in the direction of their axis, and would constantly preserve their mutual parallelisms.

Multiple bows. Corona formed.

"While these appearances are manifested new bows are formed, either commencing in the same diffuse manner or with vivid and ready formed rays; they succeed each other, passing through nearly the same phases, and arrange themselves at certain distances from each other. As many as nine have been counted having their ends supported on the earth, and in their arrangement resembling the short curtains suspended one behind the other over the scene of a theatre, and intended to represent the sky. Sometimes the intervals between these bows diminish, and two or more of them close upon each other, forming one large zone traversing the heavens and disappearing towards the south, becoming rapidly feeble after passing the zenith. But sometimes also, when this zone extends over the summit of the firmament from east to west, the mass of rays appear suddenly to come from the south, and to form, with those from the north, the real boreal corona, all the rays of which converge to the zenith. This appearance of a crown, therefore, is doubtless the mere effect of perspective; and an observer placed at the same instant at a certain distance to the north or to the south would perceive only an arc.

"The total zone, measuring less in the direction north and south than in the direction east and west, since it often leans upon the corona, would be expected to have an elliptical form; but that does not always happen: it has been seen circular, the unequal rays not extending to a greater distance than from eight to twelve degrees from the zenith, while at other times they reach the horizon.

"Let it then be imagined that all these vivid rays of light issue forth with splendour, subject to continual and sudden variations in their length and brightness; that these beautiful red and green tints colour them at intervals; that waves of light undulate over them; that currents of light succeed each other; and in fine, that the vast firmament presents one immense and magnificent dome of light, reposing on the snow-covered base supplied by the ground, which itself serves as a dazzling frame for a sea calm and black as a pitchy lake. And some idea, though an imperfect one, may be obtained of the splendid spectacle which presents itself to him who witnesses the Aurora from the Bay of Alten.

Duration of corona.

"The corona when it is formed only lasts for some minutes; it sometimes forms suddenly, without any previous bow. There are rarely more than two on the same night, and many of the Auroras are attended with no crown at all.

Disappearance of Aurora.

"The corona becomes gradually faint, the whole phenomenon being to the south of the zenith, forming bows gradually paler and generally disappearing before they reach the southern horizon. All this most commonly takes place in the first half of the night, after which the Aurora appears to have lost its intensity; the pencils of rays, the bands, and the fragments of bows appear and disappear at intervals. Then the rays become more and more diffused, and ultimately merge into the vague and feeble light which is spread over the heavens, grouped like little clouds, and designated by the name of auroral plates (plaques aurorales). Their milky light frequently undergoes striking changes in the brightness, like motions of dilatation and contraction, which are propagated reciprocally between the centre and the circumference, like those which are observed in marine animals called Medusæ. The phenomena become gradually more faint, and generally disappear altogether on the appearance of twilight. Sometimes, however, the Aurora continues after the commencement of daybreak, when the light is so strong that a printed book may be read. It then disappears, sometimes suddenly; but it often happens that, as the daylight augments, the Aurora becomes gradually vague and undefined, takes a whitish colour, and is ultimately so mingled with the cirro-stratus clouds that it is impossible to distinguish it from them."

Lieutenant Weyprecht has grandly described forms of Aurora in Payer's 'New

Lands within the Arctic Circle' (vol. i. p. 328 *et seq.*) as follows:—

Lieut. Weyprecht's description. Formation of arches.

"There in the south, low on the horizon, stands a faint arch of light. It looks as it were the upper limit of a dark segment of a circle; but the stars, which shine through it in undiminished brilliancy, convince us that the darkness of the segment is a delusion produced by contrast. Gradually the arch of light grows in intensity and rises to the zenith. It is perfectly regular; its two ends almost touch the horizon, and advance to the east and west in proportion as the arch rises. No beams are to be discovered in it, but the whole consists of an almost uniform light of a delicious tender colour. It is transparent white with a shade of light green, not unlike the pale green of a young plant which germinates in the dark. The light of the moon appears yellow contrasted with this tender colour, so pleasing to the eye and so indescribable in words, a colour which nature appears to have given only to the Polar Regions by way of compensation. The arch is broad, thrice the breadth, perhaps, of the rainbow, and its distinctly marked edges are strongly defined on the profound darkness of the Arctic heavens. The stars shine through it with undiminished brilliancy. The arch mounts higher and higher. An air of repose seems spread over the whole phenomenon; here and there only a wave of light rolls slowly from one side to the other. It begins to grow clear over the ice; some of its groups are discernible. The arch is still distant from the zenith, a second detaches itself from the dark segment, and this is gradually succeeded by others. All now rise towards the zenith; the first passes beyond it, then sinks slowly towards the northern horizon, and as it sinks loses its intensity. Arches of light are now stretched over the whole heavens; seven are apparent at the same time on the sky, though of inferior intensity. The lower they sink towards the north the paler they grow, till at last they utterly fade away. Often they all return over the zenith, and become extinct just as they came.

Band of light appears. Second band and rays.

"It is seldom, however, that an Aurora runs a course so calm and so regular. The typical dark segment, which we see in treatises on the subject, in most cases does not exist. A thin bank of clouds lies on the horizon. The upper edge is illuminated; out of it is developed a band of light, which expands, increases in intensity of colour, and rises to the zenith. The colour is the same as in the arch, but the intensity of the colour is stronger. The colours of the band change in a never-ceasing play, but place and form remain unaltered. The band is broad, and its intense pale green stands out with wonderful beauty on the dark background. Now the band is twisted into many convolutions, but the innermost folds are still to be seen distinctly through the others. Waves of light continually undulate rapidly through its whole extent, sometimes from right to left, sometimes from left to right. Then, again, it rolls itself up in graceful folds. It seems almost as if breezes high in the air played and sported with the broad flaming streamers, the ends of which are lost far off on the horizon. The light grows in intensity, the waves of light follow each other more rapidly, prismatic colours appear on the upper and lower edge of the band, the brilliant white of the centre is enclosed between narrow stripes of red and green. Out of one band have now grown two. The upper continually approaches the zenith, rays begin to shoot forth from it towards a point near the zenith to which the south pole of the magnetic needle, freely suspended, points.

Corona formed.

"The band has nearly reached it, and now begins a brilliant play of rays lasting for a short time, the central point of which is the magnetic pole—a sign of the intimate connexion of the whole phenomenon with the magnetic forces of the earth. Round the magnetic pole short rays flash and flare on all sides, prismatic colours are discernible on all their edges, longer and shorter rays alternate with each other, waves of light roll round it as a centre. What we see is the auroral corona, and it is almost always seen when a band passes over the magnetic pole. This peculiar phenomenon lasts but a short time. The band now lies on the northern side of the firmament, gradually it sinks, and pales as it sinks; it returns again to the south to change and play as before. So it goes on for hours, the Aurora incessantly changes place, form, and intensity. It often entirely disappears for a short time, only to appear again suddenly, without the observers clearly perceiving how it came and where it went; simply, it is there.

Single-rayed band.

"But the band is often seen in a perfectly different form. Frequently it consists of single rays, which, standing close together, point in an almost parallel direction towards the magnetic pole. These become more intensely bright with each successive wave of light; hence each ray appears to flash and dart continually, and their green and red edges dance up and down as the waves of light run through them. Often, again, the rays extend through the whole length of the band, and reach almost up to the magnetic pole. These are sharply marked, but lighter in colour than the band itself, and in this particular form they are at some distance from each other. Their colour is yellow, and it seems as if thousands of slender threads of gold were stretched across the firmament. A glorious veil of transparent light is spread over the starry heavens; the threads of light with which this veil is woven are distinctly marked on the dark background; its lower border is a broad intensely white band, edged with green and red, which twists and turns in constant motion. A violet-coloured auroral vapour is often seen simultaneously on different parts of the sky.

Aurora in stormy weather. Fragments.

"Or, again, there has been tempestuous weather, and it is now, let us suppose, passing away. Below, on the ice, the wind has fallen; but the clouds are still driving rapidly across the sky, so that in the upper regions its force is not yet laid. Over the ice it becomes somewhat clear; behind the clouds appears an Aurora amid the darkness of the night. Stars twinkle here and there; through the opening of the clouds we see the dark firmament, and the rays of the Aurora chasing one another towards the zenith. The heavy clouds disperse, mist-like masses drive on before the wind. Fragments of the northern lights are strewn on every side: it seems as if the storm had torn the Aurora bands to tatters, and was driving them hither and thither across the sky. These threads change form and place with incredible rapidity. Here is one! lo, it is gone! Scarcely has it vanished before it appears again in another place. Through these fragments drive the waves of light: one moment they are scarcely visible, in the next they shine with intense brilliancy. But their light is no longer that glorious pale green; it is a dull yellow. It is often difficult to distinguish what is Aurora and what is vapour; the illuminated mists as they fly past are scarcely distinguishable from the auroral vapour which comes and goes on every side.

Bands. Rays reach the pole. No noise.

"But, again, another form. Bands of every possible form and intensity have been driving over the heavens. It is now eight o'clock at night, the hour of the greatest intensity of the northern lights. For a moment some bundles of rays only are to be seen in the sky. In the south a faint, scarcely visible band lies close to the horizon. All at once it rises rapidly, and spreads east and west. The waves of light begin to dart and shoot, some rays mount towards the zenith. For a short time it remains stationary, then suddenly springs to life. The waves of light drive violently from east to west, the edges assume a deep red and green colour, and dance up and down. The rays shoot up more rapidly, they become shorter; all rise together and approach nearer and nearer to the magnetic pole. It looks as if there were a race among the rays, and that each aspired to reach the pole first. And now the point is reached, and they shoot out on every side, to the north and the south, to the east and the west. Do the rays shoot from above downwards, or from below upwards? Who can distinguish? From the centre issues a sea of flames: is that sea red, white, or green? Who can say? It is all three colours at the same moment! The rays reach almost to the horizon: the whole sky is in flames. Nature displays before us such an exhibition of fireworks as transcends the powers of imagination to conceive. Involuntarily we listen; such a spectacle must, we think, be accompanied with sound. But unbroken stillness prevails; not the least sound strikes on the ear. Once more it becomes clear over the ice, and the whole phenomenon has disappeared with the same inconceivable rapidity with which it came, and gloomy night has again stretched her dark veil over everything. This was the Aurora of the coming storm — the Aurora in its fullest splendour. No pencil can draw it, no colours can paint it, and no words can describe it in all its magnificence."

A reproduction of the woodcut in Payer's 'Austrian Arctic Voyages,' illustrating some of the features of the above description, will be found on Plate I.

In the 'Edinburgh Encyclopædia,' article "Aurora," we find:—

Descriptions of Auroræ in high Northern latitudes.

"In high Northern latitudes the Auroræ Boreales are singularly resplendent, and even terrific.

"They frequently occupy the whole heavens, and, according to the testimony of some, eclipse the splendour of stars, planets, and moon, and even of the sun itself.

In Siberia.

"In the south-eastern districts of Siberia, according to the description of Gmelin, cited and translated by Dr. Blagden (Phil. Trans. vol. lxxiv. p. 228), the Aurora is described to begin with single bright pillars, rising in the north, and almost at the same time in the north-east, which, gradually increasing, comprehend a large space of the heavens, rush about from place to place with incredible velocity, and finally almost cover the whole sky up to the zenith, and produce an appearance as if a vast tent were expanded in the heavens, glittering with gold, rubies, and sapphires. A more beautiful spectacle cannot be painted; but whoever should see such a northern light for the first time could not behold it without terror."

Maupertius's description at Oswer-Zornea.

Maupertius describes a remarkable Aurora he saw at Oswer-Zornea on the 18th December, 1876. An extensive region of the heavens towards the south appeared tinged of so lively a red that the whole constellation of Orion seemed as if dyed in blood. The light was for some time fixed, but soon became movable, and, after having successively assumed all the tints of violet and blue, it formed a dome of which the summit nearly approached the zenith in the south-west.

Red Auroræ rare in Lapland.

Maupertius adds that he observed only two of the red northern lights in Lapland, and that they are of very rare occurrence in that country.

The observations of Carl Bock, the Norwegian naturalist, kindly communicated by him to me, and detailed in Chapter III., quite confirm this observation of Maupertius as to the rare occurrence of red Auroræ in Lapland, he having only seen one.

Plate I.

THE AURORA DURING THE ICE-PRESSURE.

(Payer's Austrian Arctic Voyages.)

CHAPTER III.

SOME SPECIFIC DESCRIPTIONS OF AURORÆ, INCLUDING RESULTS OF THE ENGLISH ARCTIC EXPEDITION, 1875-76.

Captain Sabine's Auroræ.

Captain Sabine's Auroræ.

Captain Sabine describes Auroræ seen at Melville Island (Parry's first voyage, January 15). Towards the southern horizon an ordinary Aurora appeared. The luminous arch broke into masses streaming in different directions, always to the east of the zenith.

Curvature of arches towards each other.

The various masses seemed to arrange themselves in two arches, one passing near the zenith and a second midway between the zenith and the horizon, both north and south, but curving towards each other. At one time a part of the arch near the zenith was bent into convolutions like a snake in motion and undulating rapidly.

Aurora seen at Sunderland, February 8th, 1817.

('Annals of Philosophy,' vol. ix. p. 250.)

Aurora seen at Sunderland, Feb. 8, 1817. Formation of corona.

It began about 7 P.M. during a strong gale from the N.W., with single bright streamers in the N. and N.W., which covered a large space and rushed about from place to place with amazing velocity, and had a fine tremulous motion, illuminating the hemisphere as much as the moon does eight or nine days from change. About 11 o'clock part of the streamers appeared as if projected south of the zenith and looked like the pillars of an immense amphitheatre, presenting a most brilliant spectacle and seeming to be in a lower region of the atmosphere, and to descend and ascend in the air for several minutes. (This appears to have been the formation of a corona.) One streamer passed over Orion, but neither increased nor diminished its splendour.

Description of Aurora by Dr. Hayes, 6th January, 1861.

Dr. Hayes's Aurora, 6th January, 1861.

'Recent Polar Voyages' contains a narrative of the voyage of Dr. Hayes, who sailed from Boston on the 6th of July, 1860, and wintered at Port Foulbe. He witnessed a remarkable display of the Aurora Borealis on the morning of the 6th January, 1861.

Development of Aurora.

The darkness was so profound as to be oppressive. Suddenly, from the rear of the black cloud which obscured the horizon, flashed a bright ray. Presently an arch of many colours fixed itself across the sky, and the Aurora gradually developed.

Rays changed to glow.

The space within the arch was filled by the black cloud; but its borders brightened steadily, though the rays discharged from it were exceeding capricious, now glaring like a vast conflagration, now beaming like the glow of a summer morn. More and more intense grew the light, until, from irregular bursts, it matured into an almost uniform sheet of radiance. Towards the end of the display its character changed. Lurid fires flung their awful portents across the sky, before which the stars seemed to recede and grow pale.

Mixed colours. Colours change. Tongues of white flame formed.

The colour of the light was chiefly red; but every tint had its turn, and sometimes two or three were mingled; blue and yellow streamers shot across the terrible glare, or, starting side by side from the wide expanse of the radiant arch, melted into each other, and flung a strange shade of emerald over the illuminated landscape. Again this green subdues and overcomes the red; then azure and orange blend in rapid flight, subtle rays of violet pierce through a broad flash of yellow, and the combined streams issue in innumerable tongues of white flame, which mount towards the zenith.

The illustration which accompanies this description in the work is reproduced on Plate II., and forcibly reminds one of the "curtains" of the Aurora described in the preceding Chapter by Mons. Lottin.

Prof. Lemström's Aurora of 1st September, 1868.

Prof. Lemström's Aurora, 1st September, 1868.

In the first Swedish Expedition, 1868, some remarkable observations were made on the appearance of luminous beams around the tops of mountains, which M. Lemström showed by the spectroscope to be of the same nature as Auroræ.

Plate II.

AURORA SEEN BY DR. HAYES, 6TH JANUARY, 1861.

(*'Recent Polar Voyages.'*)

Aurora from earth's surface. Yellow-green line seen.

On the 1st September, 1868, on the Isle of Amsterdam in the Bay of Sweerenberg, there was a light fall of snow, and the snowflakes were observed falling obliquely. All at once there appeared a luminous phenomenon which, starting from the earth's surface, shot up vertically, cutting the direction of the falling snowflakes, and this appearance lasted for some seconds. On examination with a spectroscope the yellow-green line was found by Lemström (but of feeble intensity) when the slit of the instrument was directed towards a roof or other object covered with snow, and even in the snow all round the observer.

Lemström's conclusions.

M. Lemström concluded that an electric discharge of an auroral nature, which could only be detected by means of the spectroscope, was taking place on the surface of the ground all around him, and that, from a distance, it would appear as a faint display of Aurora.

[It should, however, here be noted that the reflection of an Aurora from a white or bright surface would give, in a fainter degree, the spectrum of the Aurora itself; and, apart from the phenomena seen by the eye, the case fails to be conclusive that an Aurora on the surface of the ground was examined.]

Mr. J. R. Capron's Aurora of October 24th, 1870.

Mr. J. R. Capron's Aurora, Oct. 24, 1870. Silver glow in north. Phosphorescent cloud-streamers. Crimson masses on horizon. Coloured streamers. Corona formed. Aurora fades away.

The description, from my notes made at the time of this fine display, is as follows:—"Last evening (October 24) the Aurora Borealis was again most beautifully seen here (Guildford). At 6 P.M. indications of the coming display were visible in the shape of a bright silver glow in the north, which contrasted strongly with the opposite dark horizon. For two hours this continued, with the addition from time to time of a crimson glow in the north-east, and of streamers of opaque-white phosphorescent cloud, shaped like horse-tails (very different from the more common transparent auroral diverging streams of light), which floated upwards and across the sky from east and west to the zenith. At about 8 o'clock the display culminated; and few observers, I should think, ever saw a more lovely sky-picture. Two patches of intense crimson light about this time massed themselves on the north-east and north-west horizon, the sky between having a bright silver glow. The crimson masses became more attenuated as they mounted upwards; and from them there suddenly ran up bars or streamers of crimson and gold light, which, as they rose, curved towards each other in the north, and, ultimately meeting, formed a glorious arch of coloured light, having at its apex an oval white luminous corona or cloud of similar character to the phosphorescent clouds previously described, but brighter. At this time the spectator appeared to be looking at the one side of a cage composed of glowing red and gold bars, which extended from the distant parts of the horizon to a point over his head. Shortly after this the Auroral display gradually faded away, and at 9 o'clock the sky was of its usual appearance, except that the ordinary tint seemed to have more of indigo, probably by contrast with the marvellous colours which had so lately shone upon it."

T. F.'s description of same at Torquay. Mr. Gibbs's report in London.

T. F., describing the same Aurora from Torquay, says it showed itself at sun-down, attained its maximum at 8, and lasted until 11. At 8 o'clock more than half the visible heavens was one sea of colour; the general ground greenish yellow and pale rose, with extensive shoals of deep rose in the east and west; while from the north, streaming upwards to and beyond the zenith, were tongues and brushes of rosy red, so deep that the sky between looked black. Mr. Gibbs reported that in London, at about 8 o'clock, brilliant crimson rays shot up to the zenith, and the sky seemed one mass of fire.

A facsimile of my water-colour sketch of this fine discharge is given on Plate III.

Mr. Barker's (superposed) red and white Auroræ, 9th November (1870?).

Mr. Barker's Auroræ, 9th November (1870?). Red Aurora. White Aurora.

On the 9th November (1870?) Mr. Barker saw at New Haven (U. S.) a most magnificent crimson Aurora. At about a quarter to 6 P.M. it consisted of a brilliant streamer shooting up from the north-western horizon. This was continued in a brilliant red, but rather nebulous, mass of light passing upwards and to the north. Its highest points were from 30° to 40° in altitude. A white Aurora, consisting of bright streamers, appeared simultaneously and extended round to the north-east. Prof. Newton informed Mr. Barker that he had observed an equally brilliant red patch of auroral light in the north-east five or ten minutes earlier.

Red seen through white.

Since the lower end of the red streamers was much lower than that of the white, it would seem as if the red were seen through the white, the red being most remote.

Crimson line not seen in white Aurora.

Spectroscopic observations of this Aurora were made. The crimson Aurora lasted less than half an hour, and then disappeared. In the white Aurora, which remained, the crimson line could not be seen.

Carl Bock's vibrating rays.

It may be here noted that during the Aurora seen by Carl Bock in Lapland, and painted by him by its own light (described, p. 25), he had the impression of sets of vibrating rays behind each other, and in the drawing it looks as if streamers were seen behind an arc.

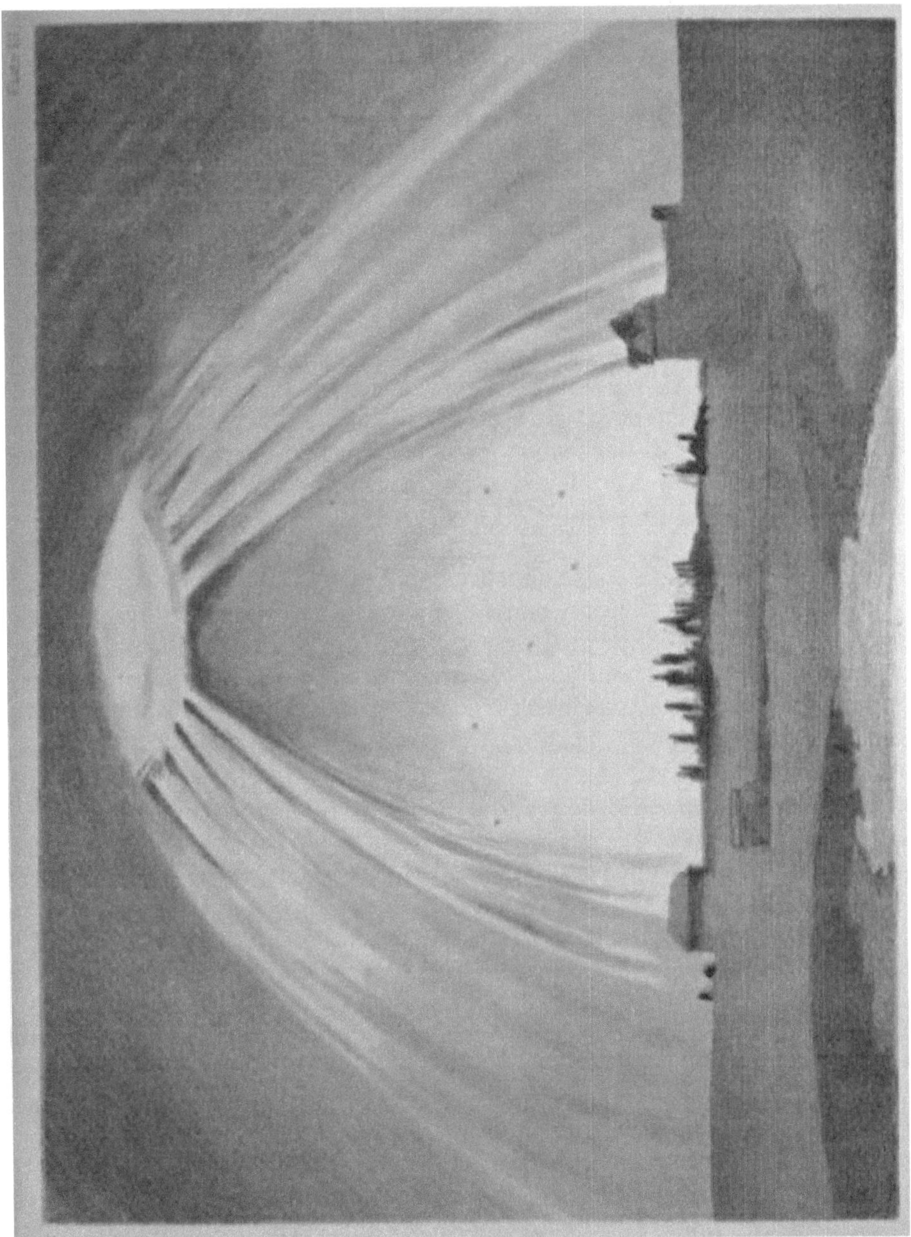

Plate III.

AURORA, GUILDFORD, OCT. 24, 1870.

From a water colour drawing.

Mr. J. R. Capron's Aurora of February 4th, 1872.

Mr. J. R. Capron's Aurora, Feb. 4, 1872. Masses of phosphorescent vapour. Rose tints appeared. Aurora from behind clouds. Formation of corona. Duration of corona. Streamers from corona. Rain during Aurora. Wind during night. Phosphorescent clouds preceded the Aurora in daylight.

My description of this Aurora as seen at Guildford, and as given at the time, is as follows:—"Last evening, returning from church a little before 8 P.M., the sky presented a weird and unusual aspect, which at once struck the eye. A lurid tinge upon the clouds which hung around suggested the reflection of a distant fire; while scattered among these, torn and broken masses of vapour, having a white and phosphorescent appearance, and quickly changing their forms, reminded me of a similar appearance preceding the great Aurora of 24th October, 1870. Shortly some of these shining white clouds or vapours partly arranged themselves in columns from east to west, and at the same time appeared the characteristic patches of rose-coloured light which are often seen in an auroral display.

From water colour drawings.

About 8 o'clock the clouds had to a certain extent broken away, and the Aurora shone out from behind heavy banks of vapour, which still rested on the eastern horizon, the north-west horizon being free from cloud and glowing brightly with red light. And now, at about 8.15, was presented a most beautiful phenomenon. While looking upwards, I saw a corona or stellar-shaped mass of white light form in the clear blue sky immediately above my head[4], not by small clouds or rays collecting, but more in the way that a cloud suddenly forms by condensation in the clear sky on a mountain top, or a crystal shoots in a transparent liquid, having too, as I thought, an almost traceable nucleus or centre, from which spear-like rays projected. From this corona in a few seconds shot forth diverging streamers of golden light, which descended to and mingled with the rosy patches of the Aurora hanging about the horizon. The spaces of sky between the streamers were of a deep purple (probably an effect of contrast). The display of the corona, though lasting a few minutes only, was equal to, if not excelling in beauty, the grand display of October 1870, before described, in which case, however, a ring or disk of white light of considerable size took the place of the stellar-shaped corona. What struck me particularly was the corona developing itself as from a centre in the clear sky, and the diverging streamers apparently shooting downwards, whereas in general the streamers are seen to shoot up from the horizon and converge overhead. The effect may have been an illusion; but, if so, it was a remarkable one. The general Aurora lasted for some time, till it was lost in a clouded sky; and, in fact, rain was descending at one time while the Aurora was quite bright. Strong wind prevailed during the night[5]. The Aurora was probably very extensive, as the evening, not-

[4] M. Lemström (Swedish Expedition, 1868) concludes that the corona of the Aurora Borealis is not entirely a phenomenon of perspective, but that the rays have a true curvature, that they are currents flowing in the same direction and attract each other. There is also an account [*antè*, p. 16] of an Aurora at Melville Island (Parry's first voyage), in which two arches were seen curving towards each other.

[5] A brilliant display in December 1870, on the east coast of Sicily, was followed by

withstanding the clouds, was nearly as bright as moonlight. The peculiar clouds referred to must have preceded the Aurora in daylight, as I recollect seeing them at 6.30 as we went to church."

AURORA. GUILDFORD. FEB. 4. 1872.
FROM A WATER COLOUR DRAWING.

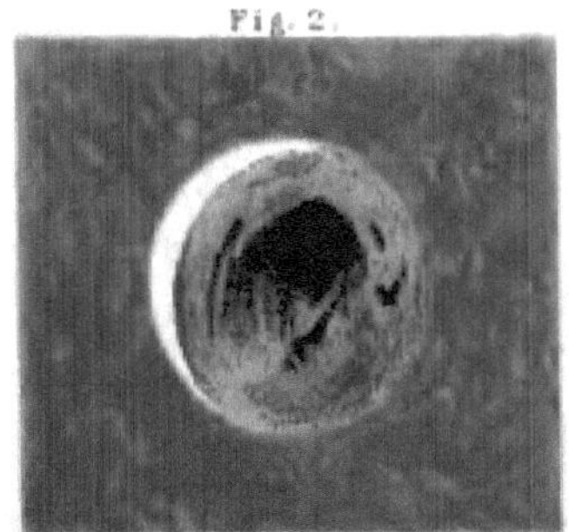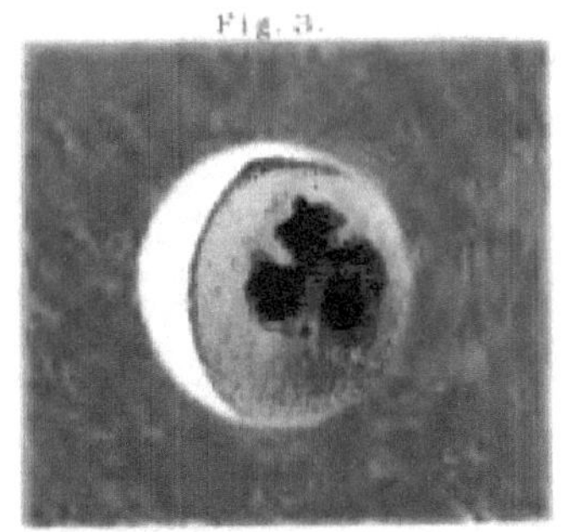

ECLIPSED MOON. GUILDFORD. AUG. 23. 24. 1877.
FROM WATER COLOUR DRAWINGS.

Plate IV.

AURORA, GUILDFORD, FEB. 4, 1872; ECLIPSED MOON, AUG. 23, 24, 1877

very violent storms, with the overflow of the Tiber and the flooding of Rome.

Aurora predicted.

They had even then a peculiarly wild, ragged, and phosphorescent appearance, and so much resembled some I had seen to accompany the Aurora of October 1870, that I predicted (as came to pass) a display later in the evening. A *facsimile* of my water-colour sketch of this Aurora is given on Plate IV. fig. 1, while the corona and rays are represented (with rather too hard an outline) on Plate V. fig. 2.

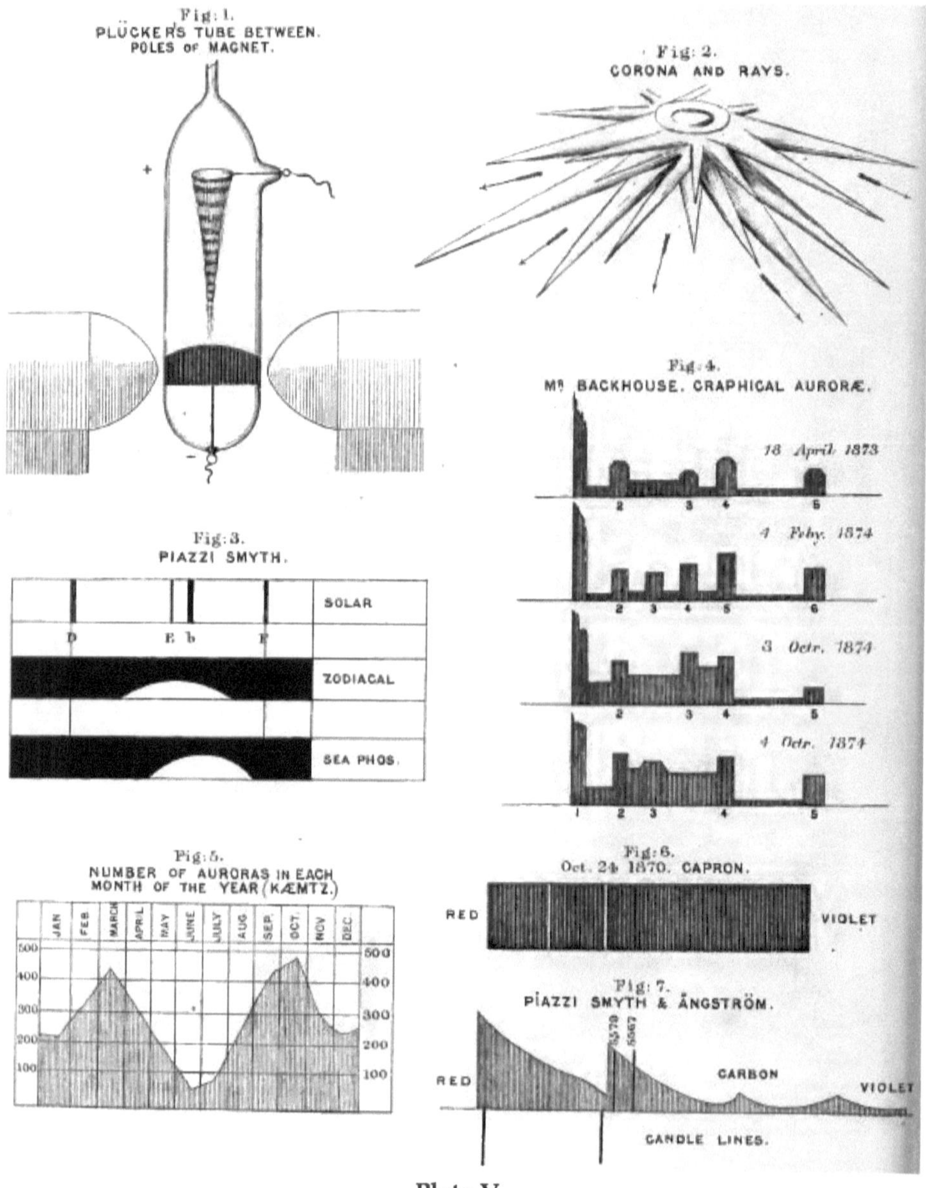

Plate V.

CORONA, GRAPHICAL AURORÆ, ZODIACAL LIGHT, &C.

Description of an Aurora seen at Cardiff.

Aurora seen at Cardiff. Formation of corona.

An Aurora was seen at Cardiff. A dusky red aspect of the sky towards the north, and extending itself across the zenith westward, made its appearance about half-past 5 P.M. The lights reached their greatest intensity at 6 o'clock, when the sky was suffused with a rich crimson glow, a broad band of colour reaching from N.E. to W. A corona of deep hue, having rugged sharply defined edges, stood out prominently in the zenith, apparently on a parallel plane to the earth, and having its centre almost immediately over the head of the spectator.

Radii thrown out from corona.

From this corona, elliptic in form, and in its broadest diameter about four times the size of the moon, there were thrown out brilliant silvery blue radii, extending to the N.E. and W. horizon, and presenting the appearance of a vast cupola of fire.

Rain fell when Aurora died out.

At half-past 6 the lights died completely out, leaving masses of cloud drifting up from the south, and a shower of rain fell. The corona was remarked upon as unusual. At Edinburgh the sky was brilliant for several hours. (The date of this Aurora is uncertain, as the account is from an undated newspaper cutting. It is supposed to be in February 1872, but could hardly have been on the 4th, as the Aurora of that date did not reach its maximum development at Edinburgh till 8 P.M.)

Mr. J. R. Capron's Aurora, seen at Guildown, Guildford, February 4th, 1874.

Silvery brightness in N.E. Light-cloud, which moved from E. to W. Formation of arc in N. Streamers. Horizontal clouds of misty light.

About 7 P.M. my attention was drawn to a silvery brightness in the north-east. Above, and still more to the east, was a bright cloud of light, which looked dense and misty, and gave one the impression of an illuminated fog-cloud. The edges were so bright that the adjacent sky, but for the stars shining in it, might, by contrast, have been taken for a dark storm-cloud. The light-cloud expanded upwards until its apex became conical, and then moved rapidly from east, along the northern horizon, until it reached the due west, where it rested, and formed for some time a luminous spot in the sky. About the same time a long low arch of light formed along the northern horizon, having a brighter patch at each extremity; and these being higher in the sky, the arch and turned-up ends were in shape like a Tartar bow. This bow was permanent; and later on a cloud of rose-coloured light formed in the east, looking like the reflection of a distant fire. From the bow also shot up curved streamers of silver light towards the zenith, which at one time threatened to form a corona. This, however, did not happen, and the Aurora gradually faded away, until, when the moon rose about 8, a silver tinge in the east alone remained. I should also mention that fleecy horizontal clouds of misty light floated in the north above the bow across the streamers.

Mr. H. Taylor informed me he saw a similar Aurora some three weeks before, in which the bright horizontal light and short white streamers were the main characters. I am not sure that the horizontal light-clouds were not actual mist-clouds illuminated by reflection of the Aurora; not so, however, I think, the first-mentioned cloud, which had more the appearance of the *aura* in the large end of an illuminated Geissler tube.

Spectrum of the Aurora described.

I examined the Aurora with a Browning direct-vision spectroscope, and found Ångström's line quite bright, and by the side of it three faint and misty bands towards the blue end of the spectrum upon a faintly illuminated ground. I could also see at times a bright line beyond the bands towards the violet. There was not light enough to take any measurements of position of the lines.

I made a pencil sketch of this Aurora, at the time when the light-cloud had moved W. and the arc formed, and of the spectrum. These drawings are reproduced on Plate VI. figs. 1 and 1*a*.

Mr. Herbert Ingall's Aurora, July 18th, 1874.

Mr. Herbert Ingall's Aurora, July 18, 1874. Haze canopy formed. Bright bluish flames appeared. Beams and streamers appeared. Oscillatory motion of rays.

An Aurora of July 18th, 1874, seen by Mr. Herbert Ingall at Champion Hill, S.E., was described by him as an extraordinary one. About 11 the sky was clear; at midnight the sky was covered by a sort of haze canopy, sometimes quite obscuring the stars, and then suddenly fading away. Mr. Ingall was shortly after remarking the sky in the S.E. and S. horizon as being more luminous than usual, when his attention was drawn to a growing brightness in the S.W., and a moment afterwards bright bluish flames "swept over the S.W. and W. horizon, as if before a high wind. They were not streamers, but bright blue flames." They lasted about a minute and faded; but about two minutes afterwards a glowing luminosity appeared in the W.S.W., and broke into brilliant beams and streamers. The extreme rays made an angle of 90° with each other, the central ray reaching an altitude of 50°. The extreme divergence of the streamers (indicating their height above the earth's surface), and their direction (from W.S.W. to E.N.E.) at right angles to the magnetic meridian, suggested to Mr. Ingall a disturbance of an abnormal character. The rays had an oscillatory motion for about fifteen seconds, and then disappeared, "as if a shutter had suddenly obscured the source of light."

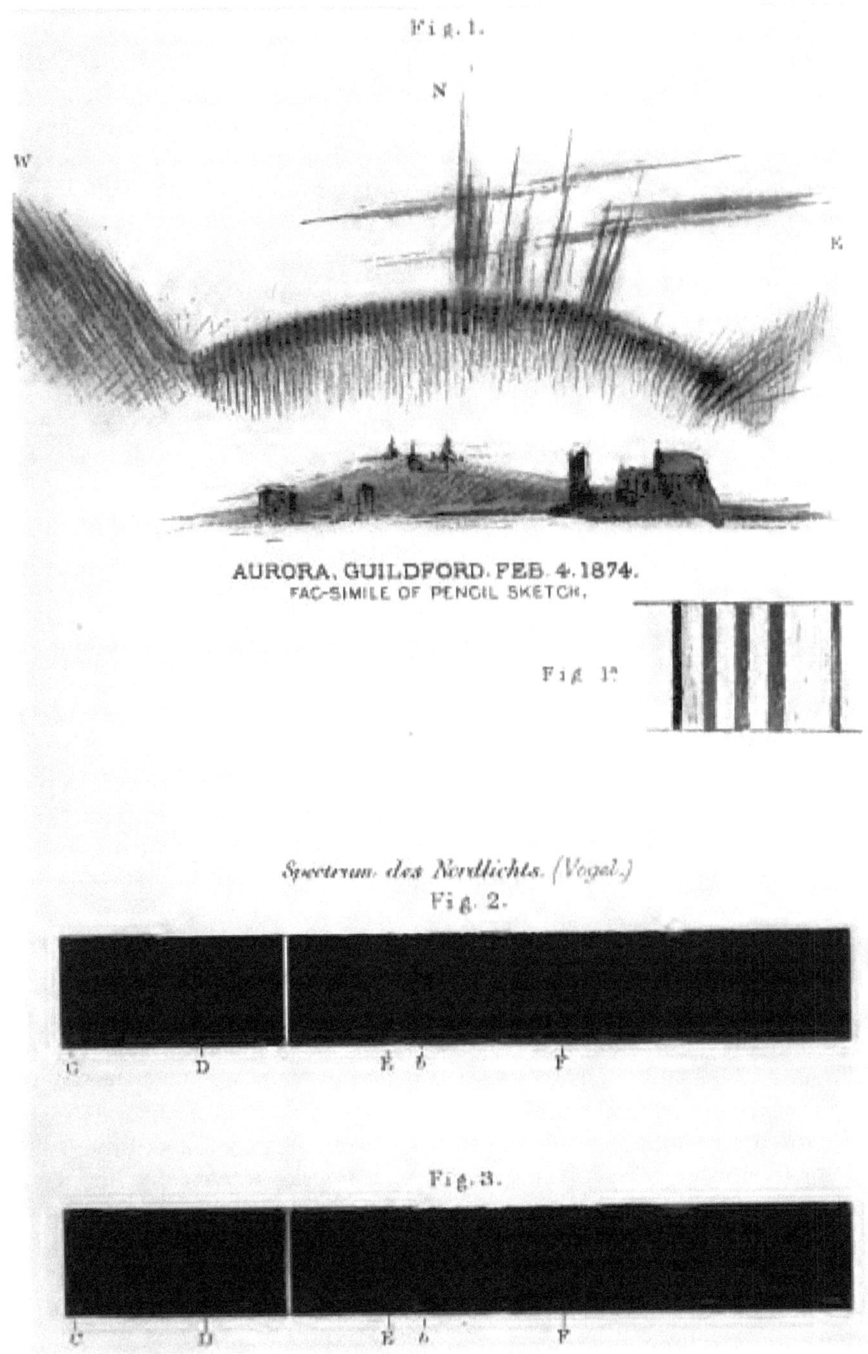

Plate VI.

AURORA, GUILDFORD, FEB. 4, 1874; SPECTRUM DES NORD-
LICHTS (VOGEL)

Mr. Ingall's remarks corroborated.

Mr. Ingall's remarks were corroborated by an observer in lat. 54° 46′ 6″·2 N., long. 6h 12m 19s·75 W. The display, however, was more brilliant, and the intensity of light at midnight illuminated the whole district as with an electric light. The rays, too, bore tints differing from one another; the largest seemed to partake of the nature of the blue sky, while the smaller ones, running parallel with the horizon, were ever changing from blue to orange-red.

Rev. C. Gape saw flashes or streaks of a pale blue colour.

On June 25 (same year?), between 9 and 10 o'clock, the Rev. Chas. Gape saw at Rushall Vicarage, Scole, Norfolk, in the E.S.E., very frequent flashes or streaks of a pale blue colour darting from the earth towards the heavens like an Aurora. The day had been dull and close, with distant thunder. In the E.S.E. it was dark, but overhead and everywhere else it was clear and starry.

Mr. J. R. Capron's White Aurora of September 11th, 1874.

Mr. J. R. Capron's white Aurora of Sept. 11, 1874.

On September 11, 1874, we were at Kyle Akin, in the Isle of Skye. The day had been wet and stormy, but towards evening the wind fell and the sky became clear. About 10 P.M. my attention was called to a beautiful Auroral display.

Double arc of pure white light in the N.

No crimson or rose tint was to be seen, but a long low-lying arc of the purest white light was formed in the north, and continued to shine with more or less brilliancy for some time. The arc appeared to be a double one, by the presence of a dark band running longitudinally through it.

White streamers. Auroral bow believed to be near the earth.

Occasional streamers of equally pure white light ran upwards from either end of the bow. The moon was only a day old, but the landscape was lighted up as if by the full moon; and the effect of Kyle Akin lighthouse, the numerous surrounding islands, and the still sea between was a true thing of beauty. The display itself formed a great contrast to the more brilliant but restless forms of Auroræ generally seen. I particularly noticed a somewhat misty and foggy look about the brilliant arc, giving it almost a solid appearance. The space of sky between the horizon and the lower edge of the arc was of a deep indigo colour, probably the effect of contrast. I had a strong impression that the bow was near to the earth, and was almost convinced that the eastern end and some fleecy clouds in which it was involved were between myself and the peaks of some distant mountains.

I have not seen any other account of this Aurora, of which I was able at the time to obtain a sketch. This is reproduced on Plate VII. It was a lovely sight, and wonderfully unlike the cloud-accompanied and crimson Auroræ which I had seen in the South.

It is noticed in Parry's 'Third Voyage' that the lower edge of the auroral arch is generally well defined and unbroken, and the sky beneath it so exactly like a dark cloud (to him often of a brownish colour), that nothing could convince to the contrary, if the stars, shining through with undiminished lustre, did not discover the deception.

No trace of brown colour in segment of sky below the arc.

I saw no trace of brown colour. The segment below the arch resting on the horizon was of a deep indigo colour.

Dr. Allnatt's Aurora, June 9th, 1876.

Dr. Allnatt's Aurora, June 9, 1876. Band of auroral light appeared. Streaks of cirro-stratus divided the Aurora. Want of electric manifestations attributed to absence of sun-spots.

Dr. Allnatt, writing to the 'Times' from Abergele, North Wales, near the coast of the Irish Channel, reported an Aurora on the night of the 9th June, 1876. After a cool and gusty day, with a strong N.E. wind and a disturbed sea, there appeared at 11 P.M. in the N. horizon a broad band of vivid auroral light, homogeneous, motionless, and without streamers. About midnight a long attenuated streak of black cirro-stratus stretched parallel with the horizon, and divided the Aurora into nearly symmetrical sections. On the preceding day the sky was covered with dark masses of electric cloud of weird and fantastic forms. The season had been singularly unproductive of high electric manifestations, which Dr. Allnatt thought might be attributable to the comparative absence of spots on the solar disk. [It may here be noted how conspicuous the years 1877 and 1878 have been for absence of Sun-spots and of Auroræ.]

Herr Carl Bock's Lapland Aurora, 3rd October, 1877.

Herr Carl Bock's Lapland Aurora, 3rd Oct. 1877. Lapland Auroræ generally of the yellow type.

In January 1878 I had the pleasure to meet, at the Westminster Aquarium, Herr Carl Bock, the Norwegian naturalist, who accompanied four Laplanders, two men and two women, with sledges, tents, &c., on their visit to this country. The Laplanders (as mentioned elsewhere) did not confirm the accounts of noises said to have been observed by Greenlanders and others during the Aurora. Carl Bock mentioned to me that the displays he saw in Lapland were most brilliant, but generally of the yellow type (the Laplanders called the Aurora "yellow lights"). He saw only one red Aurora. He kindly lent me a picture (probably in its way unique), an oil-painting of an Aurora Borealis, entirely sketched by the light of the Aurora itself.

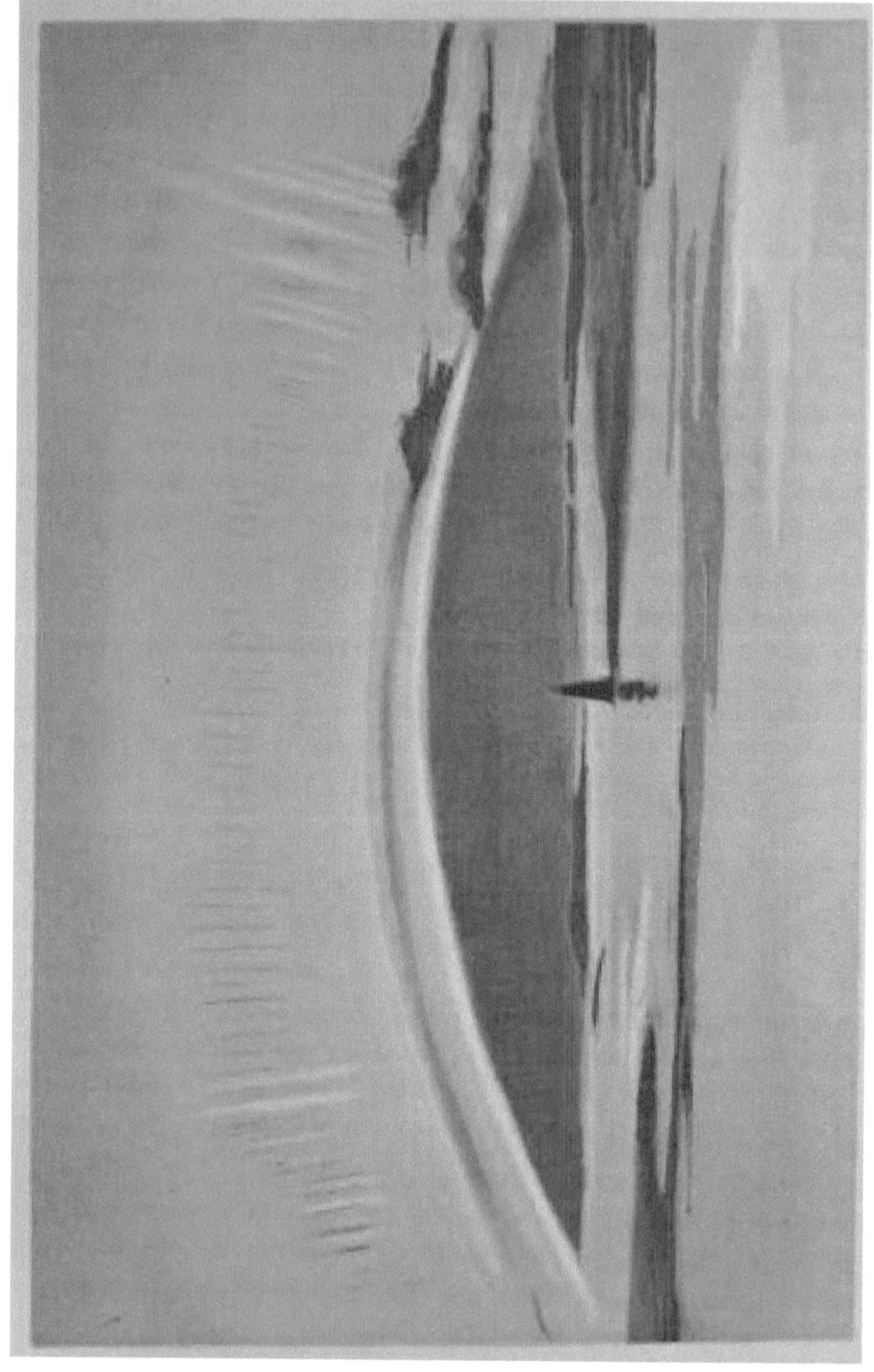

Plate VII.

AURORA, KYLE AKIN, ISLE OF SKYE, SEPT. 11, 1874

A picture painted by light of the Aurora. Movement of the rays. Inner edge of arc fringed with rays.

The painting is remarkable for the tender green of the sky, an effect probably due to a mixture of the ordinary sky colour with the yellow light of the Aurora. This picture was taken at Porsanger Fjord, in lat. 71° 50′, on 3rd October, 1877. It lasted from 9 P.M. till about 11 P.M. The rays kept continually moving, and certain of them seemed in perspective and behind the others. It will be noticed that the *inner* edge of the arc is fringed with rays, contrary to the sharp and definite margin which is usually presented. Probably two Auroræ or auroral forms were seen—a quiescent arc in front, and a set of moving streamers beyond. Two larger and brighter patches of light are seen at each extremity of the arc, as in the case of the Aurora seen by me at Guildown, February 4th, 1874, which, indeed, the display much resembles. A reduced facsimile of Herr Bock's excellent picture is given on Plate VIII.

Aurora of longitudinal rays.

Herr Bock also acquainted me that on the following day he saw an Aurora in which the lines of light, instead of being vertical, were longitudinal, and were continually swept along in several currents. They were not so strong as in the former case.

Rev. T. W. Webb's Aurora.

Rev. T. W. Webb's Aurora. Arc resolved into sets of streamers moving in opposite directions.

The Rev. T. W. Webb has described to me in a letter an Aurora very like that seen by Carl Bock in Lapland, and apparently the prevailing type in those regions. An arc similar to that figured by Carl Bock appeared in the N.W., and seemed to resolve itself into two sets of streamers moving in opposite directions (or the one set might be fixed and the other moving), like the edges of two great revolving toothed wheels. This lasted but for a few seconds; but during that interval the tints were varied and brilliant, including blue and green.

Plate VIII.
HERR CARL BOCK'S LAPLAND AURORA, OCT. 3, 1877

The English Arctic Expedition 1875-76, under Capt. Sir George Nares.

English Arctic Expedition, 1875-76. Instructions for use of officers. Appendix B. Capt. Sir G. Nares's report. True Auroræ seldom observed, and displays faint. Citron-line observed on only two displays. Appendix C.

In anticipation of the starting of this Expedition, some instructions for the use of the officers in connexion with the hoped for display of brilliant Auroræ were prepared:—as to general features of the Auroræ, by Professor Stokes; as to Polarization, by Dr. William Spottiswoode; and as to Spectrum work, by Mr. Norman Lockyer and myself. As these instructions were somewhat elaborate, and will apply to all Auroral displays, I have supplied a copy of them in Appendix B. They were unfortunately not brought into requisition, for want of the Auroræ themselves. Capt. Sir George Nares has reported to the Admiralty, under date 5th December, 1877, as follows:—"Although the auroral glow was observed on several occasions between 25 October, 1875, and 26 February, 1876, true Auroræ were seldom observed; and the displays were so faint, and lasted so short a time, and the spectrum observations led to such poor results, that no special report has been considered necessary. Although the citron-line was observed occasionally, on only two displays of the Aurora was it well defined, and then for so short a time that no measure could be obtained." (For Sir George Nares's further Report see Appendix C, containing extracts from blue-book, 'Results derived from the Arctic Expedition, 1875-6.')

Aurora Australis.

Aurora Australis. Mr. Forster's description. Long columns of white light spreading over the whole sky.

In an article on Auroræ in high Southern latitudes (Phil. Trans. No. 461, and vol. liv. No. 53), we find that Mr. Forster, who as naturalist accompanied Capt. Cook on his second voyage round the world, says:—"On February 17th, 1773, in south latitude 58°, a beautiful phenomenon was observed during the preceding night, which appeared again this, and several following nights. It consisted of long columns of a clear white light shooting up from the horizon to the eastward almost to the zenith, and gradually spreading over the whole southern part of the sky. These columns were sometimes bent sideways at their upper extremities; and though in most respects similar to the northern lights of our hemisphere, yet differed from them in being always of a whitish colour, whereas ours assume various tints, especially those of a fiery and purple hue. The sky was generally clear when they appeared, and the air sharp and cold, the thermometer standing at the freezing point." This account agrees very closely in particulars with Capt. Maclear's notice of Aurora Australis [after referred to], and especially in the marked absence of red Auroræ.

The height of the barometer does not appear to be mentioned, the temperature being apparently much the same as in the more recent cases.

Capt. Maclear's Aurora Australis, 3rd March, 1874. Light of pale yellow tint only.

In a letter dated from H.M.S. 'Challenger,' North Atlantic, April 10th, 1876, Capt. Maclear was good enough to communicate to me some particulars of an Aurora Australis seen 3rd March 1874, in lat. 54° S., long. 108° E. The letter is mainly descriptive of the spectrum (which will be described in connexion with the general question of the spectrum of the Aurora). It states that the red line was looked for in vain, and that the light appeared of a *pale yellow*, and had none of the rosy tint seen in the northern displays.

Capt. Maclear's Auroræ described in 'Nature.'

Capt. Maclear has since contributed to 'Nature,' of 1st November 1877, a description of four Auroræ seen from the 'Challenger' in high southern latitudes (including the one communicated to me). He speaks of the opportunity of observing as not frequent, either from the rarity of the phenomena, or because the dense masses of cloud prevalent in those regions prevented their being seen except when exceptionally bright. There were four appearances described:—

Feb. 9, 1874.

(1.) At 1.30 on the morning of February 9th, 1874, preceded by a watery sunset, lat. 57° S. and long. 75° E., bar. 29·0 in., ther. 35°; brilliant streaks to the westward. Day broke afterwards with high cirrus clouds and clear horizon.

Feb. 21, 1874.

(2.) At 9.30 P.M., February 21, 1874, lat. 64° S., long. 89° E., bar. 28·8 in., ther. 31°; one bright curved streamer. The Aurora preceded a fine morning with cumulo-stratus clouds, extending from Jupiter (which appeared to be near the focus) through Orion and almost as far beyond. Under this a black cloud, with stars visible through it. Real cumuli hid great part of the remainder of the sky, but there were two vertical flashing rays which moved slowly to the right (west). Generally the Aurora was still bright.

March 3, 1874. Auroral line found in light to southward.

(3.) At midnight, March 3rd, 1874, lat. 53° 30′ S., long. 109° E., bar. 29·1, ther. 36°, after some days' stormy weather, a brilliant sunset, followed by a fine morning. Soon after 8 P.M. the sky began to clear and the moon shone out. Noticing the light to the southward to be particularly bright, Capt. Maclear applied the spectroscope, and found the distinguishing auroral line.

Brilliant white clouds seen.

About midnight the sky was almost clear, but south were two or three brilliant light clouds, colour very white-yellow, shape cumulo-stratus. From about west to near south extended a long feathery light of the same colour, parallel with the horizon, and between south and west there appeared occasionally brilliant small clouds. The upper edges seemed hairy, and gave one the idea of a bright light behind a cloud. The forms changed, but no particular order was noticed.

(Here follows a description of the spectrum, and the mode in which a delineation by the lines was obtained.)

March 6, 1874. Capt. Maclear suggests whether a low barometer has to do with the absence of red.

(4.) At 8 P.M., March 6th, 1874. This was a slight Aurora, seen to the southward; after this the clouds changed to high cirrus. Capt. Maclear suggests whether a low barometer has any thing to do with the absence of red in the spectrum, the normal state of the barometer being an inch lower in those regions than in more temperate latitudes.

Barometer falls after the Aurora, and strong gale from the S. or S.W. follows.

Edin. Encyc. vol. iii. article "Aurora." Dr. Kirwan observed that the barometer commonly falls after the Aurora. Mr. Winn, in the seventy-third volume of the Phil. Trans., makes the same remark, and says that in twenty-three instances, without fail, a strong gale from the south or south-west followed the appearance of an Aurora. If the Aurora were bright, the gale came on within twenty-four hours, but was of no long continuance; if the light was faint and dull, the gale was less violent, longer in coming, and longer in duration.

Pale yellow glow rare in the Aurora Borealis.

The pale yellow-coloured glow referred to by Capt. Maclear is, in my experience, rare in the Aurora Borealis. It is probably the "æqualiter et sine eruptionibus aut radiis *fulvi*," described by Seneca (*antè*, p. 1), and may probably belong to more southern climes.

Spectrum of Auroræ Australes extends more into the violet.

We shall see too, by-and-by, that these Auroræ Australes as to spectrum extend more into the violet than the Aurora Borealis. The yellow, as complementary to violet, is likely thus to make (in the absence of the red) its appearance.

It is, however, somewhat singular that Carl Bock found almost exclusively yellow Auroræ in Lapland.

In Proctor's 'Borderland of Science,' article "The Antarctic Regions," we find quoted a passage from a letter by Capt. Howes, of the 'Southern Cross,' in which a graphic description is given of a Southern Aurora:—

Capt. Howes's description of a Southern Aurora.

"At about half-past one on the 2nd of last September the rare phenomenon of the Aurora Australis manifested itself in a most magnificent manner. Our ship was off Cape Horn, in a violent gale, plunging furiously into a heavy sea, flooding her decks, and sometimes burying her whole bows beneath the waves. The heavens were as black as death, not a star was to be seen, when the brilliant spectacle first appeared.

Balls of electric fire resting on mast-heads &c.

"I cannot describe the awful grandeur of the scene; the heavens gradually changed from murky blackness till they became like vivid fire, reflecting a lurid glowing brilliancy over every thing. The ocean appeared like a sea of vermilion lashed into fury by the storm, the waves dashing furiously over our side, ever and anon rushed to leeward in crimson torrents. Our whole ship—sails, spars, and all—seemed to partake of the same ruddy hues. They were as if lighted up by some terrible conflagration. Taking all together—the howling, shrieking storm, the noble ship plunging fearlessly beneath the crimson-crested ways, the furious squalls of hail, snow, and sleet, drifting over the vessel, and falling to leeward in ruddy showers, the mysterious balls of electric fire resting on our mast-heads, yard-arms, &c., and, above all, the awful sublimity of the heavens, through which coruscations of auroral light would shoot in spiral streaks, and with meteoric brilliancy,—there was presented a scene of grandeur surpassing the wildest dreams of fancy."

The foregoing picture presents a singular contrast to the yellow-white Auroræ described as seen in high southern latitudes by Capt. Maclear, and is interesting as a southern Aurora of a red or ruddy tint. Looking, however, at the extreme rarity of red Auroræ in those latitudes, and the description of "mysterious balls of electric fire resting on our mast-heads, yard-arms, &c." (a phenomenon not often noticed in connexion with the Aurora), it suggests itself that the case in question may have been an instance not of a true Aurora, but of an electric display, with conditions approaching those experienced by travellers who have found themselves in mountainous districts surrounded by storm-clouds charged with electricity[6].

Prof. Piazzi Smyth's Typical Auroræ.

Prof. Piazzi Smyth's typical Auroræ.

Prof. Piazzi Smyth was kind enough lately to send me the fourteenth volume of the 'Astronomical Observations made at the Royal Observatory, Edinburgh, during the years 1870-1877.' This volume, amongst its other interesting matter, affords some valuable information on the subject of the Aurora Borealis. The Aurora plates are five in number, three comprising some well-executed chromo-lithographs of typical Auroræ, from sketches made by Prof. Smyth, the other two plates being of the Aurora spectrum. The Auroræ delineated are thus described:—

Aug. 6, 1871, quiescent arc. August 21, 1871, active arc.

Plate 5. (August 6, 1871.) An example of a mild quiescent kind of auroral arc, with dark cavernous substratum. (August 21, 1871.) An example of a bright large active arc darting out rays.

---

[6] Some curious instances have been recently (January 1879) given in the 'Times' of such electric phenomena, comprising, amongst others, gas lighted by the finger in Canada, points of flame seen on the ironwork of Teignmouth Bridge, and similar points seen on the alpenstocks and axes of a party making a mountain ascent in Switzerland.

Sept. 7, 1871, arc streamers and clouds. May 8, 1871, double arc (longitudinal).

Plate 6. (September 7, 1871.) An auroral arc, with streamers and dark clouds, and maintaining a bright appearance though in proximity to the moon. (May 8, 1871.) A double-arched auroral arc (the arches are longitudinally arranged).

April 28, 1871, multiple arc. Oct. 25, 1870, coloured Aurora.

Plate 7. (April 28, 1871.) A multiple-arched arc of Aurora with moonlight. (October 25, 1870.) A case of grandest coloured Auroræ, or Aurora superb and almost universal.

All the foregoing drawings are very vivid and striking, and form a most interesting set of typical forms of Auroræ.

According to my own experience, the Aurora with arches arranged longitudinally, thus, ⌒⌒, is the rarest of all the forms. I have not met with it myself, nor do I recollect an illustration of one other than Prof. Smyth's.

CHAPTER IV.

PHENOMENA SIMULATING AURORÆ.

Auroric Lights (Kinahan).

Mr. Kinahan's Auroric Lights. White and red. White light appears in pencils radiating from a point.

Mr. G. Henry Kinahan writes to 'Nature,' from Ovoca, under date January 27th, 1877, and speaks of two distinct kinds of light so classed—one brilliant and transparent, of a white yellowish-blue or yellowish-red colour, while the other is semi-opaque and of a bloody red colour, the latter being considered in Ireland a forerunner of bad weather. The first kind generally appears as intermittent pencils of light that suddenly appear and disappear.

Frequently not stationary, but jumping about.

Usually they proceed or radiate from some point near the north of the horizon; but Mr. Kinahan has frequently seen them break from a point in the heavens, not stationary, but jumping about within certain limits. Sometimes these lights occur as suddenly flashing clouds of light of a white colour, but at other times of blue and reddish yellow.

In daylight like sun-rays. Red light appears in clouds floating upwards or diffused.

If this class of lights is watched into daylight, they appear somewhat like faint rays of a rising sun. One morning, while travelling in West Galway in the twilight, they were very brilliant, and quite frightened Mr. Kinahan's car-driver, who thought the sun was going to rise in the north instead of the east. The second, or bloody red light, usually occurs in clouds floating in one direction up into the heavens, but often diffused over a portion of the sky. Mr. Kinahan has never seen them coming from the east, and on only a few occasions from the south, but generally from the west, north-west, or north.

Red light appears as dirty misty clouds in daylight, or as a mist or misty rays.

If both kinds of light appear at the same time, the second while passing over the first dims it. If the second class is watched into daylight, they appear as dirty misty clouds that suddenly form and disappear without the spectator being able to say where they come from or where they go to, or as a hazy mist over a portion of the sky, that suddenly appears and disappears, or as misty rays proceeding from a point in the horizon. Generally, when these clouds occur, there is a bank of black clouds to the westward.

Season since October 1876 prolific in auroric light.

Mr. Kinahan then speaks of the season as having been prolific in auroric light, as there had been few nights since the 1st October then last (1876) in which they did not appear. On many occasions they were late in the night, being very common and brilliant during the dark days of December, a few hours before dawn (about 5 o'clock). Each time there was a fine day they appeared also, and the weather broke again.

Mr. J. Allan Broun questions nature of these lights, as Aurora is seldom seen at 5 A.M. in this country. On 77 occasions seen only twice so early. Season was of marked infrequency elsewhere.

Mr. Jno. Allan Broun refers to this graphic account of Mr. Kinahan's, and concludes there must have been some mistake as to the nature of these "auroric lights," as the Aurora Borealis is very rarely seen at 5 A.M. in this country. In the years 1844 and 1845, during which the Aurora was sought for at Makerstown every hour of the night, it was observed in 77 nights on an average of nearly three hours each night; but it was seen only twice so early, and that with a bright or brilliant Aurora, which remained during five hours on the first occasion, and from 6 P.M. to 6 A.M. on the second. Parts of the phenomenon seen by Mr. Kinahan, Mr. Broun also could not say he had ever seen; and if Mr. Kinahan's observations could have been confirmed it would have been most important, especially as made so frequently at the epoch of minimum. The description is in many respects a sufficiently recognizable one of auroral discharges; but the frequent appearance in early morning is certainly unusual, and few if any Auroræ seem to have been recorded as appearing elsewhere in Great Britain during the time which Mr. Kinahan refers to as so prolific (see, however, Dr. Allnatt's, *antè*, p. 24). In fact, the season in question was one of marked infrequency (see English Arctic Expedition Report, *antè*, p. 26). Mr. Buchan furnished Mr. Broun with a note of Auroræ seen in the stations of the Scottish Meteorological Society during the year 1876, and they were 42 in number, 26 in the first half, and 16 in the second half of the year. The greater part were seen in the most northerly stations, including the Orkney, Shetland, and Faroe Islands, and only 9 south of the Forth.

Luminous Arch.

Luminous arch, Sept. 11, 1814. Height above horizon 6 to 9 miles.

In the 'Annals of Philosophy,' vol. iv. p. 362, there is a minute description of a luminous arch which appeared in the sky on the night of Sunday, September 11th, 1814, and was seen in the west of England opposite the Irish Sea, the west part of the south of Scotland, and part of the west of Ireland. It was described as a part of either a body of dense greyish-white light, or a mass of luminous matter in the shape of an arch. Its height above the horizontal line was estimated at not more than 9 nor less than 6 miles.

It moved southward, and was assumed to differ from the Aurora.

Its direction when first seen was N. 80° E., and S. 80° W. It moved to the southward. It was assumed to differ from the Aurora Borealis in wanting coruscations, and in its having a much paler light.

CHAPTER V.

SOME QUALITIES OF THE AURORA.

Noises attending Auroræ.

In the Edinb. Encyc., Gmelin is stated, in continuation of his description of an Arctic Aurora, to add:—"For however fine the illumination may be, it is attended, as I have heard from the relation of many persons, with such a hissing, cracking, and rushing noise through the air, as if the largest fireworks were playing off." To describe what they then heard, the natives are said to use the expression, "Spolo-chi chodjat"—that is, The raging host is passing. The hunter's dogs, too, are also described as so much frightened when the Auroræ overtake the hunters, that they will not move, but lie obstinately on the ground till the noise has passed. This account of noises seems to be confirmed by other testimony. They are stated to have been heard at Hudson's Bay and in Sweden; and Musschenbroek mentions that the Greenland whale-fishers assured him they had frequently heard the noise of the Aurora Borealis, but adds that "no person in Holland had ever experienced this phenomenon." Mr. Cavallo declares he "has repeatedly heard a crackling sound proceeding from the Aurora Borealis" (Elements of Nat. or Exper. Phil. vol. iii. p. 449). Mr. Nairne mentions that in Northampton, when the northern lights were very bright, he is confident he perceived a hissing or whizzing sound. Mr. Belknap of Dover, New Hampshire, North America, testifies to a similar fact (American Trans. vol. ii. p. 196).

Sir John Franklin negatives them.

Sir John Franklin mentions, in his 'Journey to the Shores of the Polar Sea':— "Nor could we distinguish its (the Aurora's) rustling noise, of which, however, such strong testimony has been given to us that no doubt can remain of the fact."

March 11th. Hissing noise heard during Aurora's passage. Explained to arise from the snow.

In detail, he mentions he never heard any sound that could be unequivocally considered as originating in the Aurora, although he had had an opportunity of observing that phenomenon for upwards of 200 nights (the Aurora was registered at Bear Lake 343 times without any sound being heard to attend its motions); but the uniform testimony of the natives and all the older residents in the country induced him to believe that its motions were sometimes audible. On the 11th March, at 10 P.M., a body of Aurora rose N.N.W.; and after a mass had passed E. by S., the remainder broke away in portions, which crossed about 40° of the sky with great rapidity. A hissing noise, like that of a bullet passing through the air, was heard, which seemed to proceed from the Aurora; but Mr. Wentzel assured the party the noise was occasioned by severe cold succeeding mild weather, and acting upon the surface of the snow previously melted in the sun's rays. A similar noise was heard the next morning.

Capt. Sabine also negatives noise.

In Parry's first voyage, Captain Sabine describes an Aurora seen at Melville Island, and adds that the Aurora had the appearance of being *very near* the party, but *no sound could be heard.*

Article "Aurora Polaris," Encyc. Brit., suggests noises as not improbable.

In the article "Aurora Polaris," Encyc. Brit, edition ix., the writer admits the evidence of scientific Arctic voyagers having listened in vain for such noises; but, referring to the statements of Greenlanders and others on the subject, concludes there is no *à priori* improbability of such sounds being occasionally heard, since a somewhat similar sound accompanies the brush-discharge of the electric machine.

Payer negatives and discredits noises.

Payer, of the Austrian Polar Expedition (1872-1874), states that the Aurora was never accompanied by noise, and discredits the alleged accounts of noises in the Shetlands and Siberia.

As also Weyprecht.

Lieut. Weyprecht, of the same expedition, says (*antè*, p. 14):—"Involuntarily we listen; such a spectacle must, we think, be accompanied with sound, but unbroken silence prevails, not the least sound strikes on the ear."

Herr Carl Bock negatives noises in the case of Lapland Auroræ.

Herr Carl Bock, who accompanied the Laplanders visiting this country (at the Westminster Aquarium) in 1877-78, and who witnessed many brilliant auroral displays in Lapland, assured me he could trace no noise, except on one occasion, when he heard a sort of rustling, which he attributed to the wind. The Laplanders themselves did not associate any special noise with the Aurora.

Auroral noises in telephone. Ringing sound in vacuum-tube under influence of magnet.

It has been recently stated, in an article on the Telephone in 'Nature,' that Professor Peirce "has observed the most curious sounds produced from a telephone in connexion with a telegraph wire during the Aurora Borealis;" but no further details are given. In experimenting with a silicic fluoride vacuum-tube between the poles of an electro-magnet, I found, on the magnet being excited, that the capillary stream of blue light was decreased in volume and brightness, and at the same time from within the tube a peculiar whistling or slightly metallic ringing sound was heard.

Adverse conclusion as to noises accompanying Aurora.

I certainly have never met with an instance of noise accompanying an Aurora and traced to it. On the whole the balance of evidence seems quite adverse to any proof of noises proper ordinarily accompanying an Aurora.

Colours of the Aurora.

Colours of the Aurora. Sir John Franklin's views. Other observers have described all colours of spectrum. Violet rare. Crimson indicates coming Aurora.

Sir John Franklin considered the colours in the Polar Aurora did not depend on the presence of any luminary, but were generated by the motion of the beams, and then only when that motion was rapid and the light brilliant. The lower extremities, he says, quivered with a fiery red colour, and the upper with orange. He also saw violet in the former. Other observers have, in their various descriptions of Auroræ, mentioned the colours of the rays or beams as red, crimson, green, yellow, &c.; in fact, comprising the range of the spectrum. Violet seems less frequently mentioned. The red or crimson colour is frequently the first indication of the coming Aurora, and is usually seen on or near the horizon. The colours have frequently been observed to shift or change.

Prof. Piazzi Smyth describes colours of Aurora of Feb. 4, 1872, as seen at Edinburgh.

Prof. Piazzi Smyth, in a letter to 'Nature,' describing the Aurora of February 4th, 1872, as seen at Edinburgh, says that when the maximum development was reached all the heavens were more or less covered with pink ascending streamers, except towards the N., which was dark and grey—first by means of a long low arch of blackness, transparent to large stars, and then by the streamers which shot up from this arch, which were green and grey only for several degrees of their height, and only became pink as they neared the zenith. The red streamers varied from orange to rose-pink, red rose, and damask rose.

The Professor pointed out that the spectroscope knew no variety of reds giving one red line only, and attributed this to the mixing up of rays and streamers of blackness out of the long low arch. When the Aurora faded away a true starlight-night sky appeared; so that evidently the dark arch and streamers were as much part of the Aurora as the green and red lights.

Dr. Allnatt at Frant describes vivid colours of same Aurora.

Dr. Allnatt, at Frant, found in the case of the same Aurora the south-western part of the heavens tinged by a bright crimson band. A dark elliptical cloud extending from S. to S.E. was illuminated at its upper edge with a pale yellow light, and sent up volumes of carmine radii interspersed with green and the black alternating matter characteristic of elemental electricity. Almost due E., and of about 25 degrees elevation, was a bright insulated spot of vivid emerald-green, which appeared almost sufficiently intense to cast a faint shadow from intercepting objects. At 7 o'clock the Aurora had passed the zenith, and the sky presented a weird and wonderful appearance. A dark rugged cloud, some 8 degrees E. of the zenith, was surrounded by electric light of all hues—carmine, green, yellow, blood-red, white, and black; and the bright spot still existed in the south.

Descriptions at Blackburn and Cambridge. Lapland Auroræ yellow.

At Blackburn, in Lancashire, the rays were described as glowing in the N.E. from silvery white to deepest crimson; and at Cambridge the same Aurora was described as of a brilliant carmine tint. The Auroræ seen in Lapland by Herr Carl Bock, were, he informed me, almost invariably yellow; he saw only one red one.

Hydrogen vacuum-tube suggestive of Aurora colours.

The behaviour of a hydrogen Geissler vacuum-tube will be subsequently referred to in the Chapter on the comparison of some tubes with the Aurora spectrum, and is suggestive as to Aurora colours.

Variation of tints in.

The capillary part of this tube, when lighted by a small coil, was found to vary in tint—silver-white, bright green, and crimson being each in succession the dominant colour, according to the working of the break of the coil. When a spectroscope was used, the red, blue, and violet lines of the gas were seen to change in intensity in accordance with the light colour seen in the tube.

Variation of colour in nitrogen tube under influence of magnet.

A Geissler nitrogen vacuum-tube was also so arranged that the capillary part of it should be vertically between the conical extremities of the armatures of a large electro-magnet, the armatures just being clear of the outside of the tube. The tube was then lighted up by a small coil, and the magnet excited by four large double-plate bichromate cells.

Change from rosy to violet hue.

The stream of light was steady and brilliant, and, except at the violet pole, of the rosy tint peculiar to a nitrogen vacuum-tube. On excitation of the electro-magnet, the discharge was seen to diminish in volume, with an apparent increase in impetuosity; and not only the capillary part, but in a less degree the bulbs also of the tube, changed from a rosy to a well-marked violet hue.

Photographic plates taken. Difference in.

We several times connected and disconnected the magnet with its batteries, but always with the same result. Of the spectrum of the capillary part of this tube we took photographic plates with quartz prisms and lenses, taking care that all things should be as equal as possible, the apparatus undisturbed, and the time of exposure exactly the same. One plate was taken with the tube in its normal condition, the other while it was under the influence of the magnet. The spectra were identical, except that the plate of the tube influenced by the magnet was decidedly the brightest, and was found to penetrate more into the violet region (the Author's 'Photographed Spectra,' p. 60, plate xxv.). These plates effectually corroborated the change of colour, as the violet ray would have more photographic effect than the rosy. The identity of the spectra of the capillary part proved that the change in colour could not have proceeded from an extension of the violet glow. (A similar experiment will be found also detailed in Part III. Chapter XII.)

Height of the Aurora.

Height of Aurora. Sir John Franklin considers it within the region of the clouds. At no great elevation.

Sir John Franklin (Narrative of a Journey on the Shores of the Polar Sea in the years 1819, '20, '21, '22) says:— "My notes upon the appearance of the Aurora coincide with those of Dr. Richardson in proving that that phenomenon is frequently seated within the region of the clouds, and that it is dependent in some degree upon the cloudy state of the atmosphere." And further:— "The observations of Dr. Richardson point particularly to the Aurora being formed at no great elevation, and that it is dependent upon certain other atmospheric phenomena, such as the formation of one or other of the various modifications of cirro-stratus."

Sir John Franklin also refers to notes from the Journal of Lieut. Robert Hood, R.N., on an Aurora:—

Observations of Lieut. Robert Hood and Dr. Richardson. A beam not more than 7 miles from the earth. An arch 7 miles from the earth.

The observations were made at Basquian House, and at the same time by Dr. Richardson at Cumberland House, quadrants and chronometers having been prepared for the purpose. On the 2nd April the altitude of a brilliant beam was 10° 0′ 0″ at 10h 1m 0s at Cumberland House. Fifty-five miles S.S.W. it was not visible. It was estimated that the beam was not more than 7 miles from the earth, and 27 from Cumberland House. On the 6th April the Aurora was for some hours in the zenith at that place, forming a confused mass of flashes and beams; and in lat. 53° 22′ 48″ N., long. 103° 7′ 17″, it appeared in the form of an arch, stationary, about 9° high, and bearing N. by E. It was therefore 7 miles from the earth.

An arch between 6 and 7 miles from the earth.

On the 7th April the Aurora was again in the zenith before 10 P.M. at Cumberland House, and in lat. 53° 36′ 40″ N., long. 102° 31′ 41″. The altitude of the highest of two concentric arches at 9h P.M. was 9°, at 9h 30m it was 11° 30′, and at 10h 0m 0s P.M. 15° 0′ 0″, its centre always bearing N. by E. During this time it was between 6 and 7 miles from the earth. [The bearings are true, not magnetic.]

Sir John Franklin's remarks.

Sir J. Franklin says this was opposed to the general opinion of meteorologists of that period: he also noticed he had sometimes seen an attenuated Aurora flashing across the sky in a single second, with a quickness of motion inconsistent with the height of 60 or 70 miles, the least that had hitherto been ascribed to it.

Dr. Richardson's conclusions.

The needle was most disturbed, February 13, 1821, P.M., at a time when the Aurora was distinctly seen passing between a stratum of cloud and the earth; and it was inferred from this and other appearances that the distance of the Aurora from the earth varied on different nights. Dr. Richardson concludes that his notes prove, independent of all theory, that the Aurora is occasionally seated in a region of the air below a species of cloud which is known to possess no altitude; and is inclined to infer that the Aurora Borealis is constantly accompanied by, or immediately precedes, the formation of one or other of the forms of cirro-stratus.

Captain Parry observed Auroræ near the Earth's surface. Sir W. R. Grove's observation at Chester. Mr. Ladd's observation at Margate. The author's observation at Kyle Akin, Skye.

Captain Parry observed Auroræ near the earth's surface; and records that he and two companions saw a bright ray of the Aurora shoot down from the general mass of light between him and the land, which was distant some 3000 yards. Sir W. R. Grove ('Correlation of Physical Forces') saw an Aurora at Chester, when the flashes appeared close, so that gleams of light continuous with the streamers were to be seen between him and the houses—"he seemed to be in the Aurora." Mr. Ladd, of Beak Street, Regent Street, has related to me an appearance he was struck with, and examined carefully. Standing in the evening in Margate Harbour, he saw a white ray of the Aurora, which, apparently shooting downwards, was clearly placed between his eye and the opposite head of the pier, which projected into the sea. Mr. Ladd also informed me that Prof. Balfour assured him that such an appearance was not unusual. In the double-arc Aurora seen by me in the Isle of Skye, September 11, 1874 (described *antè*, p. 23), I had a strong impression that the bow was near the earth, and thought that the eastern end, and some fleecy clouds in which it was involved, were between myself and the peaks of the distant mountains.

Dalton's calculation of 100 miles. Backhouse's 50 to 100 miles. Prof. Newton, mean 130 miles. Elevation of Auroræ cannot exceed a few miles.

In the article "Aurora Polaris," Encyc. Brit., edition ix., Dalton is instanced as having calculated the height of an Aurora in the north of England at 100 miles; and Backhouse as having made many calculations, with the result of an average height of 50 to 100 miles. Prof. Newton, too, is quoted for the height of 28 Auroræ (calculated by one observation of altitude and amplitude of an arch) as ranging from 33 to 281 miles, with a mean of 130 miles. It is, however, pointed out that a height of 62 miles above the earth's surface would imply a vacuum attainable with difficulty, even with the Sprengel pump. This difficulty is then met by a reference to the observed altitude of some meteors, and to a suggestion of Prof. Herschel's that electric repulsion may carry air or other matter up to a great height. Dr. Lardner ('Museum of Science and Art,' vol. x. p. 192) speaks of the height of Auroræ as not certainly ascertained; but considers them atmospheric phenomena scarcely above the region of the clouds, and does not think it probable that their elevation in any case can exceed a few miles.

M'Clintock's observations. Capt. Ross saw Auroræ on an ice-cliff, which he attributed to electric action.

M'Clintock, after noticing that the beams of the Aurora were most frequently seen in the direction of open water, says that in some cases patches of light could be plainly seen a few feet above a small mass of vapour over an opening in the ice. Captain Ross, in his Antarctic voyage, saw the bright line of the Aurora forming a range of vertical beams along the top of an ice-cliff; and suggested this was produced by electrical action taking place between the vaporous mist thrown upwards by the waves against the berg, and the colder atmosphere with which the latter was surrounded.

Bergman estimates height as 468 miles.

Bergman, from a mean of 30 computations, makes the height of the phenomenon to be 72 Swedish (about 468 English) miles.

Boscovich 825 miles.

Father Boscovich calculated the height of an Aurora Borealis observed on the 16th December, 1727, to have been 825 miles.

Mairan 600 miles. Euler several thousand miles. Dr. Blagden about 100 miles.

Mairan supposed the far greater number of Auroræ to be at least 600 miles above the surface of the earth. Euler assigned them an elevation of several thousands of miles. Dr. Blagden, however, limited their height to about 100 miles, which he supposed to be the region of fireballs—remarking that instances were upon record in which northern lights had been seen to join and form luminous balls, darting about with great velocity, and even leaving a train behind them like common meteors (Phil. Trans. vol. lxxiv. p. 227).

Dalton 150 miles.

Mr. Dalton, from an observation of the luminous arches on a base of 22 miles, found the altitude of the Aurora to be about 150 miles (Dalton's 'Meteorological Observations and Essays,' 1793, pp. 54, 153).

Dr. Thompson assumes considerable height. His table. Average of 31 observations, 500 miles.

Dr. Thompson, 'Annals of Philosophy,' vol. iv. p. 429 (1814), assumes that the height of the beams above the surface of the earth was much greater than that of most other meteorological appearances, and gives (p. 430) a table of Auroræ, mainly taken from Bergman, Opusc. v. p. 291, of 31 Auroræ observed in the years 1621 to 1793, with heights in English miles. The lowest is, 23rd February, 1784, London (Cavendish), 62 miles; the highest, 23rd October, 1751, Fournerius, 1006 miles! The average of the 31 estimated observations gives a height of about 500 miles. It is not stated how these observations were obtained, though methods are mentioned how they might be.

Prof. Heis's instrument for determining height of Auroræ.

Prof. Heis, of Münster, exhibited at the recent Scientific Loan Collection at South Kensington ('Official Catalogue,' 3rd edit. p. 296, No. 1231) an instrument for the determination of the position of the point of convergence of the rays of the Aurora, and for determining the height of the Aurora. A ball resting in a pan was to be brought into position, so that several diverging pencils of Aurora, when properly viewed, were covered by the rod which passed through the centre of the ball. The point of the rod (which could be moved up and down in the ball), when the instrument was set to the astronomical meridian, showed the azimuth and altitude of the converging point of the pencils of light. This point of convergence does not coincide with the point to which the inclination-needle directs. From the deviation of the two points, the height of the Aurora could be calculated.

Professor Newton's method of calculating height.

Professor H. A. Newton (Sil. Journ. of Science, 2nd ser. vol. xxix. p. 286) has proposed a method of calculating the height of Auroræ by one observation of altitude and amplitude of an arch. It assumes that the auroral arches are arcs of circles, of which the centre is the magnetic axis of the earth, or at least that they are nearly parallel to the earth's surface, and probably also to the narrow belt or ring surrounding the magnetic and astronomical poles. Professor Newton finds that, d being the distance from the observer to the centre of curvature of the nearest part of this belt (for England, situated about 75° N. lat., 50° W. long.), h the apparent altitude of the arch, $2a$ its amplitude on the horizon, x its height, R the earth's radius, and c the distance of the observer from the ends of the arch:—

$$\sin \varphi = \sin d \cos a \operatorname{cosec}(d + h) \qquad (1)$$

$$\tan c = z \sin h \sin \varphi \sec{}^2\varphi \qquad (2)$$

$$x = R - (\sec c - 1) \qquad (3)$$

Gave a height from 33 to 281 miles, and a mean of 130 miles.

This method with 28 Auroræ gave a height from 33 to 281 miles and a mean of 130 miles.

Galle has suggested (Pogg. Ann. cxlvi. p. 133) that the height of Auroræ might be calculated from the amount of divergence between the apparent altitude of the auroral corona and that indicated by the dipping-needle, a principle which has been adopted in Prof. Heis's apparatus before described. The results do not differ materially from Professor Newton's.

The conclusions to be arrived at from the foregoing instances and opinions are certainly very puzzling. The terrestrial character of some Auroræ seems well established. The height to which these phenomena *may* ascend is left almost a matter of conjecture, and further observations are very desirable.

Phosphorescence.

Phosphorescence. Phosphorescent bands. Storm-clouds which threw out cirri. Shone with a sort of phosphorescence. Storm-cloud surrounded by glories of a phosphorescent whiteness.

In the voyage of the 'Hansa' ('Recent Polar Voyages,' p. 420), on the 9th September, 1869, at 10 P.M., Aurora gleams appeared in the west, shooting towards the south. "Radiant sheaves and phosphorescent bands mounted towards the zenith," but the phantasmagoria quickly vanished. M. Silbermann ('Comptes Rendus,' lxviii. p. 1120) mentions storm-clouds which threw out tufts of cirri from their tops, which extended over the sky, and resolved into, first, fine, and afterwards more abundant rain. (I saw a fine day example of this on the Lago di Guarda, ending in a copious discharge of rain attended with loud thunder and vivid lightning.) Usually the fibres were sinuous; but in much rarer cases they became perfectly rectilinear and surrounded the cloud like a glory, and occasionally shone *with a sort of phosphorescence.* On the night of 6th September, 1865, at 11 P.M., a stormy cloud was observed in the N.N.W., and lightning was seen in the dark cumulous mass. Around this mass extended *glories of a phosphorescent whiteness*, which melted away into the darkness of the starry sky. Round the cloud was a corona, and outside this two fainter coronæ. After the cloud had sunk below the horizon the glories were still visible.

Sabine's luminous cloud at Loch Scavaig, Skye. Other observations of luminous clouds.

Sabine mentions a cloud frequently enveloping Loch Scavaig, in Skye, as being at night perfectly self-luminous, and that he saw rays, similar to those of the Aurora, but produced in the cloud itself. Sabine also refers to luminous clouds mentioned in Gilbert's Annals, and to observations by Beccaria, Deluc, the Abbé Rozier, Nicholson, and Colla; and to luminous mists as observed by Dr. Verdeil at Lausanne in 1753, and by Dr. Robinson in Ireland.

Aurora at Melville Island.

He also describes (Parry's First Voyage) an Aurora seen at Melville Island, and says the light was estimated as equal to that of the moon when a week old. Besides the pale light, *which resembled the combustion of phosphorus*, a slight tinge of red was noticed when the Aurora was most vivid; but no other colours. This Aurora was repeatedly seen *on the following day.*

Procter suspects Aurora is formed in a mist. M'Clintock: Aurora is never visible in a perfectly clear atmosphere.

Mr Procter, in a letter to me, suspects that the Aurora is generally formed in a sort of "mist or imperfect vapour;" and this mist or imperfect vapour seems in many instances to form part of the Aurora, and to partake of its self-luminous character. M'Clintock does not imagine that the Aurora is ever visible in a perfectly clear atmosphere. He has often observed it just silvering or rendering luminous the upper edge of low fog or cloud-banks, and with a few vertical rays feebly vibrating.

Aurora of Feb. 4, 1874. Illuminated fog-cloud. Capt. Oliver's meteor-cloud. Auroral display, 24th Oct., 1870. Streamers of phosphorescent cloud.

An instance of apparent phosphorescence is supplied by the Aurora of the 4th February, 1874 (*antè*), when a bright cloud of light was seen which gave the impression of an *"illuminated fog-cloud."* Captain S. P. Oliver saw at Buncrana, Co. Donegal, on February 4, 1874, what he describes as a meteor-cloud, viz. "a broad band of silvery white and luminous cloud." This appearance, as described by another correspondent, was evidently an imperfectly formed (perhaps actually forming) Auroral arc. The great Auroral display of the 24th of October, 1870, as seen by me, included, according to my notes made at the time, "streamers of opaque white phosphorescent cloud, very different from the more common transparent Auroral diverging streams of light."

Aurora of Feb. 4, 1872, at Frant. Radii of phosphorescent light.

Describing the Aurora of February 4, 1872, at Frant, Dr. Allnatt says:—"At a later hour of the night the canopy of cirro-stratus had separated, and was transformed into luminous masses of radiant cumulus. At 10.40 the Aurora reappeared in the N., and sent luminous radii of white *phosphorescent* light from the periphery of a segment of a perfectly circular arch"[7].

The author's description of same Aurora. Masses of phosphorescent vapour.

Again, February 4th, 1872, as described by me, the first signs of the Aurora were (in dull daylight) a lurid tinge upon the clouds, which suggested the reflection of a distant fire; while scattered among these, "torn and broken masses of white vapour having a phosphorescent appearance" reminded me of a similar observation in October 1870.

Day Auroræ must have a phosphorescent glow. Ångström considers yellow-green line due to fluorescence or phosphorescence. Oxygen and some of its compounds phosphorescent.

The day Auroræ, which are elsewhere described, and are not very uncommon, could, we may presume, hardly be seen without the presence of some phosphorescent glow. Professor Ångström, in his Aurora Memoir (discussed elsewhere), in discussing the yellow-green line, considers the only probable explanation to be that it owes its origin to fluorescence or phosphorescence. He says that some fluorescence is produced by the ultra-violet rays; and adds, "an electric discharge

[7] On the occasion of the Aurora of September 24, 1870, Dr. Allnatt says, "the air seemed literally alive with the unwonted phosphorescence."

may easily be imagined, which, though in itself of feeble light, may be rich in ultra-violet light, and therefore in a condition to cause a sufficiently strong fluorescent light." And he refers to the fact that oxygen and some of its compounds are phosphorescent.

A phosphoretted hydrogen spectrum-band is close to yellow-green auroral line. Phosphorescent or fluorescent after-glow of electric discharge.

In the examination of certain spectra connected with the Aurora, detailed in Part II., I have shown that the bright edge of one of the phosphoretted hydrogen bands is in close proximity to the yellow-green Auroral line. I have also referred to the peculiar brightening by reduction of temperature of one of the bands in the red end of the spectrum of phosphoretted hydrogen, so that from almost invisible it became bright, and to the peculiar brightening of a line in the yellow-green in certain "Aurora" and phosphorescent tubes. It has also been observed that the electric discharge has a phosphorescent or fluorescent after-glow (isolated, I believe, by Faraday). It seems difficult to avoid in some way connecting all these circumstances with the yellow-green line of the Aurora, if not also with the line in the red.

Sorby's experiments on fluorescence and absorption. Bonelleine, spectrum of. Coloured layer of fungi. Spectrum of Oscillatoriæ.

Mr. Sorby, in his experiments on the connexion between fluorescence and absorption ('Monthly Microscopical Journal'), found in the spectrum of a solution in alcohol of a strongly fluorescent substance called bonelleine (the green colouring-matter found in the *Aurelia Bonellia-viridis*) two bright bands, the one red and the other green, with centres respectively at 6430 and 5880, and their limits towards the blue end at 6320 and 5820. On adding an acid the red band changed its place to 6140. The superficial membranous coloured layer of the fungi *Russula nitida* and *vesca* in alcohol gave an absorption band with centre at 5540, while the spectrum of fluorescence extended to 4400. A solution of *Oscillatoriæ* in water gave a spectrum of absorption with bands at 6200 and 5690; while the spectrum of fluorescence showed two bright bands having their centres at 6470 and 5800, and their limits towards the blue end at 6320 and 5710.

Sea phosphorescence, a continuous spectrum.

These instances of course cannot be connected with the Aurora except as showing the spectrum region and lines of fluorescence. The sea phosphorescence, according to Professor Piazzi Smyth, has a continuous spectrum extending from somewhat below E to near F (Plate V. fig. 3).

Ångström finds the sky almost phosphorescent.

Ångström, on the occasion of the starry night when he found traces of the green line in all parts of the heavens, speaks of the sky as being "almost phosphorescent."

Author of article in Encyc. Brit. suggests that the phosphorescent or fluorescent light may be due to chemical action. Herschel's observation of phosphorescence in Geissler and "garland" tubes.

The author of the Aurora article in the Encyc. Brit. suggests that the phosphorescent or fluorescent light attributed to the Aurora may be due to chemical action. He also questions Ångström's assumption that water-vapour is absent in the higher atmosphere, and thinks that it and other bodies may, by electric repulsion, be carried above the level they would attain by gravity. He then continues that if discharges take place between the small sensible particles of water or ice in the form of cirri (as Silbermann has shown to be likely) surface decomposition would ensue, and it is highly probable the nascent gases would combine with emission of light. He adds "that it has been almost proved that in the case of hydrogen phosphide the very characteristic spectrum (light?) produced by its combustion is due neither to the elements nor to the products of combustion, but to some peculiar action at the instant of combination; and it is quite possible that under such circumstances as above described water might also give an entirely new spectrum." Professor Herschel has referred to the phosphorescent light which remains glowing in Geissler tubes after the spark has passed, and to the fact that one of the globes of a "garland" tube which was heated did not shine after the spark had passed, apparently because of the action of heat on the ozone to which the phosphorescence might be due. (See experiments on Mr. Browning's bulbed tube, Part III. Chap. XV.)

Aurora and Ozone.

Aurora and Ozone. Smells of sulphur during Auroræ attributed to ozone.

Accounts are given by travellers in Norway of their being enveloped in the Aurora, and perceiving a strong smell of sulphur, which was attributed to the presence of ozone. M. Paul Rollier, the aëronaut, descended on a mountain in Norway 1300 metres high, and saw brilliant rays of the Aurora across a thin mist which glowed with a remarkable light. To his astonishment, an incomprehensible muttering caught his ear; when this ceased he perceived a very strong smell of sulphur, almost suffocating him ('Arctic Manual,' p. 726).

Question whether the oxygen of the air may be changed into ozone.

In the case of the Aurora, the question naturally arises whether the oxygen of the air may be changed into ozone, perhaps also whether the nitrogen may not be modified in some similar manner.

Ozone destroyed by heat.

The absorption spectra of oxygen, and of the same gas in its form of ozone, may possibly differ; but this can hardly happen in the case of incandescent oxygen, for ozone is at once destroyed by heat at 300°, and slowly at 100°, and must be partially at least destroyed by the heat of the discharge. If any lines were due to ozone in such a spectrum, we should expect they would be weakened by heat and brightened by cold.

Ozone in a large bell-receiver not manifested in spectrum.

In the case of a continued discharge in a large exhausted bell-receiver, the presence of ozone in considerable quantities was manifested to us by its odour when the receiver was removed from the pump; but the spectrum of the stream of light did not appear to differ from that in Geissler tubes.

Professor Dewar demonstrates that ozone is condensed oxygen.

In a course of lectures at the Royal Institution in March 1878, on the Chemistry of the Organic World, Prof. Dewar appears to have demonstrated, by Prof. Andrews' apparatus, that ozone is really condensed oxygen, and, further, that during this condensation heat is absorbed, which is evolved during the decomposition or re-expansion.

Refers to the silent discharge between the atmosphere and the earth.

He also exhibited the oxidizing power of ozone in its action on mercury, and commented on its similar action upon organic matter in forming nitrates, and on its remarkable bleaching properties, but added there was as yet no proof of its combining with free nitrogen. That peroxide of hydrogen accompanies the formation of ozone by the slow combustion of phosphorus, and that this peroxide acts with ozone in decomposing organic bodies, though in an inexplicable manner, the Professor considered to be proved. He also referred to the silent discharge probably perpetually going on between the upper and lower strata of the atmosphere, and also between these and the earth, accounting, as the Professor considered, for some of the chemical actions whereby nitrogenous compounds are formed in the soil.

No spectrum of ozone obtained.

As far as I am aware, no information as to a possible spectrum of ozone, or a modification of the oxygen or other spectra by its presence, has, up to the present time, been obtained[8].

Suggestion to subject electric discharge to influence of cold.

It has been suggested by Mr. Procter and myself that the electric discharge in an exhausted moist tube, if subjected to a considerable degree of cold, might produce a modification of the air-spectrum, perhaps even a spectrum analogous to that of the Aurora.

For some further notes on this subject see Appendix D (Aurora and Ozone).

Polarization of the Aurora Light.

Polarization of the Aurora light. Mr. Ranyard found none.

In 'Nature,' vol. vii. p. 201, is contained an account of observations of the po-

---

[8] See, however, Dr. Schuster's article "On the Spectra of Lightning," Phil. Mag. May 1879, p. 316.

larization of the zodiacal light and of the Aurora, by Mr. A. Cowper Ranyard, who, using both a double-image prism and a Savart on the great Aurora of February 4th, 1872, detected no trace of polarization. He also examined a smaller one of 10th November, 1871, with a like result.

Prof. Alexander found strong polarization in latitude 60°.

Mr. Fleming (who refers to these observations) remarks that the only other account he had met with was contained in Prof. Stephen Alexander's Report on his Expedition to Labrador, given in Appendix 21 of the U.S. Coast Survey Report for 1860, p. 30. Professor Alexander found strong polarization with a Savart's polariscope, and thought that the dark parts of the Aurora gave the strongest polarization. This was in latitude about 60°, at the beginning of July, and near midnight. It is not stated whether there was twilight or air-polarization at the time, nor is the plane of polarization given.

Mr. Shroeder found no polarization.

The question naturally arises, especially as the darkest parts of the Aurora are usually situated low down near the horizon, whether the polarization in the latter case did not proceed from the atmosphere and not from the Aurora itself. Mr. Shroeder found no traces of polarization in the Aurora of February 4th, 1872. Further examinations of the Aurora with some delicate form of polariscope would seem very desirable.

Polarization not found in the zodiacal light; except faint traces by Mr. Burton.

The evidence of polarization in the case of the zodiacal light seems also almost entirely negative—Mr. Ranyard pointing out observations of his own, of Captain Tupman, and of Mr. Lockyer with this result. Mr. Burton, using a Savart set so as to give a black centre when the bands were parallel to the plane of polarization, believed he detected faint traces of polarization in the brightest parts of the zodiacal light (as seen in Sicily), the bands being black-centred when their direction coincided with the axis of the cone of light. Mr. Burton saw no trace of bands when examining the slight remaining twilight apart from the zodiacal light. Mr. Ranyard was not able to confirm Mr. Burton's observations on the same evening and with the same instrument.

Number of Auroræ.

Number of Auroræ. Sir John Franklin's observations.

Sir John Franklin saw in the Arctic Regions, in the years 1819, 1820, 1821, 1822:—In the month of September two Auroræ, in October three, in November three, in December two, in January five, in February seven, in March sixteen, in April fifteen, and in May eleven.

Periodicity as to days not established.

Periodicity as to days seems to have no certain law; and though certain days in February and March are marked as those of fine returning displays, they must be looked on as accidental.

Maxima and minima.

Two well-marked annual maxima seem to occur in March and October (the latter the greater), and two minima in June and January, the greater in June (Encyc. Brit.). The 4th of February, 1872, and same day 1874, are, however, curious instances of a recurring remarkable display.

Kæmtz's table.

A table by Kæmtz, showing the number of Auroras in each month of the year, with the maxima and minima as above stated, will be found on Plate V. fig. 5.

Dr. Hayes's observations in winter of 1860-61.

Dr. Hayes has observed that in the winter of 1860-61 (when the ten or eleven years' inequality was at its maximum) only three Auroræ were seen and recorded, and they were feeble and short in duration.

Captain Maguire's observations at Point Barrow as to number and time of appearances.

Captain Maguire, at Point Barrow (1852-54), reports that the Aurora was seen six days out of seven, and on 1079 occasions, being nearly one third of the hourly observations. It was seldom seen between 9 A.M. and 5 P.M., not at all between 10 A.M. and 4 P.M. It increased regularly and rapidly from 5 P.M. until 1 A.M., and then diminished in the same way until 9 A.M.

The winters of 1877 and 1878 and the springs of 1878 and 1879 have been singularly deficient in Auroræ. I have seen none at Guildown.

Duration of Aurora.

Duration of Aurora. Sometimes a few minutes; at other times the whole night or even days.

In the article in the 'Edinb. Encyc.' before referred to some remarks are made on the duration of the Aurora. Sometimes it is formed and disappears in the course of a few minutes. At other times it lasts for hours or during the whole night, or even for two or three days together. Musschenbroek observed one in 1734 which he considered to have lasted ten days and nights successively, and another in 1735 which lasted from the 22nd to the 31st March.

Auroræ may run on into the day without being noticed.

With respect to Captain Maguire's observations (*antè*) it may be remarked that Auroræ may doubtless frequently run on into and through the day without their being noticed (instances, however, are known of Auroræ seen in daylight); and hence it is difficult to judge of the limit of duration of a particular Aurora unless indications are sought for during the day (by the shapes of clouds, action of the magnet, &c.) as well as during the night. Probably Auroræ seen during successive nights may be parts of a continuous discharge.

The Travelling of Auroræ.

Travelling of Auroræ. Donati's investigations.

Donati undertook to study the Aurora with reference to the mode of its extension; and he arrived at the result that the Aurora of February 4, 1872, was not observed in different regions of the earth in the same physical moment; *but everywhere at the same local hour*, as in the case of celestial phenomena, which do not share in the earth's rotation.

Questions sent to Italian Consuls.

The Minister of Foreign Affairs sent a circular to all Italian Consuls, asking them the necessary questions; and in reply received reports from forty-two places in our hemisphere and from four in the southern, the places embracing in one latitude the considerable extent of 240 degrees of longitude.

An epitome of the tables (in which the results are divided into three zones) is as follows:—

Table of results.

Zone.	Mean longitude of zone.	No. of stations.	Mean hour of maximum.	Mean hour of end.
Eastern	2 hrs. 5 mins. E.	9	9½ hrs.	12¼ hrs.
Middle	0 hr. 20 mins. E.	17	8½ hrs.	11½ hrs.
Western	5 hrs. 38 mins. W.	13	8¾ hrs.	9¾ hrs.

Extensions of the Aurora. The Aurora passed through four periods. First period of origin, light weak. Second period, increase of intensity. Third period, continuous brightness. Fourth period, decrease.

Donati summed up the facts:—That the light phenomena of this Aurora began to show themselves in the extreme east of the southern hemisphere in Eden and Melbourne; shortly after, they were observed in the east of our hemisphere in China (but not in Japan); from China the Aurora passed over the whole of Asia and Europe, and crossed the Atlantic and the American Continent as far as California. It was invisible in Central and South America. During these immense extensions it passed through four periods. In the first (called by Donati the period of origin) the light of the Aurora was pretty weak, and spread from Shanghai to Bombay; in the second period, during which it passed on from Bombay to Taganrog, it acquired a sudden increase of intensity; in the third period (called by Donati the normal) the Aurora passed over Europe from east to west with regularity and a continuous brightness; the fourth period, that of decrease, was observed in America. The Aurora had a tendency to end earlier in reference to the local hour in the western stations than in the eastern. The acceleration on an average of the end of the phenomenon was twenty minutes for every hour of longitude.

Donati's conclusions. Explanation of mode of propagation of same Aurora.

Donati concluded that these facts were not reconcilable with the theory of the Aurora depending on meteorological and electro-magnetic phenomena of the globe. Since, too, we have not a yearly, but a ten-yearly period of the Aurora, which coincides with that of sun-spots and terrestrial magnetism, Donati supposed that the cosmic causes of the polar lights were electro-magnetic currents between the sun and the earth. This would explain the mode of propagation of the Aurora of 4th February. Conceive an electric current going from the earth to the sun, or *vice versâ*; certain phenomena of the Aurora could only be observed in those parts of the atmosphere which have a determinate position or direction with reference to this current; and consequently these phenomena would be successively visible on the different meridians, as these meridians, by reason of the earth's rotation, assume the same position to the current. For the Aurora to be visible certain meteorological and telluric circumstances must, however, doubtless work together with the cosmical cause.

Geographical Distribution of Auroræ (Fritz and Loomis).

Geographical Distribution of Auroræ. Prof. Fritz's and Prof. Loomis's line of frequency.

Professors Fritz and Loomis have investigated this subject; and Petermann's 'Mittheilungen,' vol. xx. (1874), contains a paper by the former, from which it appears that the northern limit of Auroræ chosen by Professor Loomis nearly coincided, except in England, with a line of frequency in Professor Fritz's paper. This line nearly passes through Toronto, Manchester, and St. Petersburg. Professor Loomis places it as far north as Edinburgh. On a line across Behring's Straits, and coming down below 60° N. in America and the Atlantic, and just north of the Hebrides, to Dröntheim, and including the most northern points of Siberia, the frequency is represented by 100.

Within this another zone of greatest frequency and intensity.

Within this is another zone of greatest frequency and intensity, which passes just south of Point Barrow, in lat. 72° N., on the northern coast of America, and by the Great Bear Lake to Hudson's Bay, where it reaches a latitude of 60°, then on to Nain, on the coast of Labrador, and to the south of Cape Farewell; then bending sharper to the northward, it passes between Iceland and the Faroe Islands, near to the North Cape, on by the northern ice-sea to Nova-Zembla and Cape Tschejuskin, and on just to the north of the Siberian coast to the south of Kellett Land, thence returning to Point Barrow.

Lines on which annually nearly the same number of Auroræ are seen.

More or less parallel with this line are the lines on which annually nearly the same number of Auroræ are seen. The line for one Aurora annually went from Bordeaux, through Switzerland, past Krakau, south of Moscow and Tobolsk, to the northern end of Lake Baikal, on to the Sea of Ochotsk and to the Southern Aleutes, thence through Northern California to the mouth of the Mississippi and to Bordeaux. The line for five Auroræ annually went from Brest through Belgium, Stettin, Wologda, between Tobolsk and Beresow, parallel to the previous line to Ochotsk, and on to Brest, &c. Almost exactly with the line of greatest frequency coincides the line forming the boundary of the direction of visibility of the Northern Light towards the Pole or towards the Equator; while northwards of this line the Polar Light is seen in the direction towards the Equator; and from all stations the Northern Lights are seen in directions which are pretty much normal to that curve and the entire system of isochasms.

Assumed connexion between Aurora and ice-formation.

Professor Fritz has remarked that the curves of greater frequency tend towards the region of atmospheric pressure, and also that they bear some relation to the limit of perpetual ice—tending most southward where, as in North America, the ice limit comes further south. He also endeavours to show a connexion between the periods of maximum of Auroræ and those of ice-formation, and considers ice to be an important local cause influencing their distribution. These being most frequently seen over open water in the Arctic regions, has been referred to as noticed by Franklin and others.

Extent and principal Zone of the Aurora.

Extent and principal zone of Aurora. M. Moberg's Finland observations (1846-55) compared by Prof. Fritz with those in other regions.

The Finland observations, published by M. Moberg in his 'Polarlichter Katalogue' of Northern Lights in the years 1846-55, numbering 1100, have been compared by Prof. Fritz, in his paper in the 'Wochenschrift für Astronomie,' with the auroral phenomena of the same period in all other regions. The Table shows that of 2035 days of the months August to April on which Northern Lights were seen, 1107 days were those of Northern Lights for Finland. On 794 they were visible simultaneously in America, and mostly also in Europe; on 101 days in Europe only, and on 212 days in Finland only. On 958 days Northern Lights were visible in Europe and America which were not visible in Finland. All these numbers refer only to the months August to April, as in the remaining months the brightness of the night in Finland makes such observations impossible.

The conclusion is arrived at that a large portion of Auroræ have no very great extension, or that the causes producing the phenomena must often be of a very local character; while in another portion of the phenomena the extent, or the regions of simultaneous appearance are very considerable.

Number limited to Finland only small.

The number limited to Finland, for which hitherto corresponding observations from other lands are wanting, is very small—212, or 19 per cent. of the whole number seen in Finland. With the increase of frequency of the phenomena at the time of maximum, the number observed in Finland and America on the same day increases; while those observed in Finland and Europe only, or in Finland only, decreased, in accordance with the known law that with the frequency the intensity and extent also increase.

One third of Auroræ seen in America and Europe simultaneously.

Between 1826 and 1855, of 2878 days on which, in America, the Northern Lights were seen, there were 1065 on which they were also visible in Europe; so that at least every third day of Auroræ was common to both these portions of the globe. In the years 1846 to 1855, and 1868 to 1872, there were in the first period 657 Northern-Light days common to America and Europe out of 1691, and in the second 397 out of 715.

Local occurrence of the Aurora not in favour of its assumed cosmical nature.

The comparison by Prof. Fritz of M. Moberg's Finland observations has been lately reviewed in 'Nature' (March 8, 1878) and the result arrived at that, "After ten years, in spite of the vastly accumulated material of careful observations, there appears no necessity to change Herr Fritz's system of curves in any essential detail; indeed certain parts of the same, which were at first only based on probability and supposition (the part of the principal zone between the north of Norway and Nishen Kolynisk as an instance), we now know with perfect certainty to be correct." It has been remarked that the local occurrence of Auroræ is not in accordance with the hypothesis of the phenomenon being one of a cosmical nature.

The winter of 1870 was remarkable for brilliant displays; and the displays of October 24th and 25th, 1870, were remarkably brilliant in England and in America also, and the Aurora Australis was seen on the same days at Madras. These displays were seen in England and America in the daytime as patches or coronæ of white light, with streamers stretching upward from them.

CHAPTER VI.

THE AURORA IN CONNEXION WITH OTHER PHE-NOMENA.

Auroræ and Clouds.

Auroræ and clouds. Dr. Richardson's observations. Aurora constantly accompanied by or immediately precedes the formation of cirro-stratus.

Dr. Richardson ('Sir John Franklin's Narrative'), so long ago as the years 1819-1822, made many recorded observations on the connexion of clouds with the Aurora Borealis in the Polar regions. Some of these are alluded to in Chapter V., section "Height of the Aurora," for the purpose of showing the moderate distance he found it to be above the earth; and his inference is there mentioned, "that the Aurora Borealis is constantly accompanied by or immediately precedes the formation of one or other of the various kinds of cirro-stratus." On the 13th November and 18th December, 1820, the connexion of an Aurora with a cloud intermediate between cirrus and cirro-stratus is mentioned. It is, however, also mentioned that the most vivid coruscations of the Aurora were observed when there were only a few attenuated shoots of cirro-stratus floating in the air, or when that cloud was so rare that its existence was only known by the production of a halo round the moon. (An instance of attenuated streaks of cirro-stratus in connexion with an auroral arc will be found in the Aurora seen at Guildown on the 4th February 1874, a sketch of which is reproduced on Plate VI. fig. 1.)

Polarity discerned in cirro-stratus clouds.

Dr. Richardson goes on to express his opinion that he, on some occasions, discerned a polarity in the masses of clouds belonging to a certain kind of cirro-stratus (approaching cirrus), by which their long diameters, having all the same direction, were made to cross the magnetic meridian nearly at right angles.

Apparent polarity of Aurora might perhaps be ascribed to the clouds themselves.

Dr. Richardson further suggests that if it should be thereafter proved that the Aurora depends upon the existence of certain clouds, its apparent polarity might perhaps be ascribed to the clouds themselves which emit the light; or, in other words, the clouds might assume their peculiar arrangement through the operation of one cause (magnetism, for instance), while the emission of light might be produced by another—a change in their internal constitution perhaps connected with a motion of the electric fluid.

Dr. Richardson further remarks that, generally speaking, the Aurora appeared in small detached masses for some time before it assumed that convergency towards the opposite parts of the horizon which produced the arched form.

Sir John Franklin's observations.

Sir John Franklin says in his Polar expeditions he often perceived the clouds in the daytime disposed in streams and arches such as the Aurora assumes.

Dr. Low's.

Dr. Low ('Nature,' iv. p. 121) considers he witnessed a complete display of auroral motions in cirrus cloud, and considers all clouds subject to magnetic or diamagnetic polarization; he states that when the lines converge towards the magnetic pole fine weather follows, and when at right angles to it wet and stormy.

M. Silbermann's observations, 15th April, 1869. Cirrus clouds took the place of the Aurora.

In the Encyc. Brit. edition 9, article "Aurora Polaris," after referring to the evidence of Franklin, Richardson, and Low, M. Silbermann ('Comptes Rendus,' lxviii. p. 1051) is quoted in detail for observed connexion between the Aurora and cirrus cloud. 15th April, 1869, at 11h 16m, an Aurora appeared and disappeared; but it seemed as if the columns were still visible, and it soon became obvious that fan-like cirrus clouds, with their point of divergence in the north, had taken the place of the Aurora. Between 1 and 2 A.M. the clouds had passed the zenith, and let fall a little fine frozen rain. At 4 A.M. the cirrus of the false Aurora was still visible, but deformed towards the top, and presenting a flaky aspect. The cirrus never appeared to replace the Aurora either from right or left, but to substitute itself for it like the changes of a dioramic view.

Payer thinks the transition of Aurora into clouds not proved.

Payer, in his 'Austrian Arctic Voyages,' thinks that the occurrence of the Aurora during the day (i. e. *light clouds with its characteristic movement*) had been rather imagined than actually observed, and that the transition of white clouds into auroral forms at night has never been satisfactorily proved. He, however, mentions the mist-like appearance of the Aurora.

Dr. Allnatt's observations, 4th February, 1872, at Frant. Aurora passed into cirro-stratus.

Dr. Allnatt observed the splendid Aurora of 4th February, 1872, at Frant, and noticed the weird and wonderful appearance of the phenomena. At 6 P.M. the Aurora commenced by the S.W. portion of the heavens being tinged with a bright carmine hue, and in a short time the whole visible hemisphere was lighted up. A dark elliptical cloud extending from S. to S.E. and S.W. sent up volumes of coloured radii. At 7 the Aurora had passed the zenith, and a dark, broken, rugged cloud some 8° E. of zenith was surrounded by electric light of all hues. At 7.40 the Aurora began to wane, and passed into a homogeneous cirro-stratus of sufficient density to obscure the stars, disappearing at 7.45.

Later, cirro-stratus was transformed into luminous cumulus.

At a later hour of the night the canopy of cirro-stratus had separated and was transformed into luminous masses of radiant cumulus; so that, as Dr. Allnatt observes, there were called in requisition almost all the most prominent cloud-modifications during the progress of the phenomena. The succession of formation, transformation, and reformation from Aurora to cloud and from cloud to Aurora was, Dr. Allnatt concluded, conclusive of the theory before advanced of the electric origin of the recurrent rayed cloud-modifications in the place of the magnetic meridian, over which so much mystery had been cast.

Aurora and Thunder-storms.

Aurora and thunder-storms. Silbermann's theory.

Silbermann asserts that Auroræ are produced by the same general phenomena as thunder-storms, and concludes that the Auroræ of 1859 and 1869 assumed the character of thunder-storms which, instead of bursting in thunder, had been drawn into the upper parts of the atmosphere, and their vapour being crystallized in tiny prisms by the intense cold, the electricity became luminous in flowing over these icy particles.

Prof. Piazzi Smyth on monthly frequency of Auroræ and storms.

Professor Piazzi Smyth has observed that the monthly frequency of Auroræ varies inversely with that of thunder-storms. His Table of comparisons is as follows:—

His table of observations.

Month.	Lightning.	Auroræ.
January	24·0	29·7
February	14·4	42·5
March	7·0	35·0
April	15·4	27·5
May	37·4	4·8
June	48·0	0·0
July	55·2	0·5
August	38·4	12·6
September	22·4	36·6
October	20·8	49·4
November	15·0	32·4

December	15·0	28·8
Mean of whole year	24·0	20·1

Silbermann's observations 15th April, 1869. 30th April, 1865.

Silbermann, on 15th April, 1869, observed a fall of rain (tiny crystals of ice) on the disappearance of an Aurora and its change into cloud forms (see section, "Auroræ and Clouds," p. 53). He also observed a rain of little sparkling ice-prisms on 30th April, 1865, at Paris, the city being then enveloped in a cirrus of vertical fibres similar to that which frequently accompanies the Aurora.

On the occasion of the Aurora seen by me at Guildown, 4th February, 1872, rain fell immediately succeeding the formation of the corona.

The falling of rain as an immediate sequence of an Aurora seems, however, to be rather the exception than the rule; but possibly this may vary with the character of the Aurora itself—whether it be of the crimson class, passing into cloud and accompanied with much electric disturbance, or of the more quiescent white.

A falling barometer observed to follow Auroræ.

A falling barometer following a display of Auroræ has been noticed by Sir John Franklin and others; and in some cases (notably one in Sicily before referred to) storms and floods have accompanied this.

Professor Christison's observations.

In a paper read before the Royal Society of Edinburgh in 1868, Prof. Christison mentioned, as a fact of importance to agriculturists, that the first great Aurora after autumn is well advanced, and following a period of fine weather, is a sign of a great storm of rain and wind in the forenoon of the second day afterwards.

Mr. C. L. Prince, in his 'Climate of Uckfield,' p. 218, remarks that displays of Auroræ are almost invariably followed by very stormy weather, after an interval of from 10 to 14 days.

Aurora and the Magnetic Needle.

Aurora and the magnetic needle. Sir John Franklin's observations. Motion communicated to the needle was neither sudden nor vibratory. Return of needle to its former position very gradual. Different positions of the Aurora had considerable influence on the direction of the needle. Needle disturbed when Aurora not visible. Quiescent yellow Aurora produced no perceptible effect on needle. Return of needle more speedy after formation of a second arch. Slow when disturbance was considerable.

Sir John Franklin, in his 'Narrative' (before referred to), gives Lieutenant Robert Hood, R.N., the credit of being "the first who satisfactorily proved, by his observations at Cumberland House (before mentioned), the important fact of the action of the Aurora upon the compass-needle," and also "to have proved the Aurora to be an electrical phenomenon, or at least that it induces a certain unusual state of electricity in the atmosphere." Sir John Franklin then mentions that the motion communicated to the needle was neither sudden nor vibratory. Sometimes it was simultaneous with the formation of arches, prolongation of beams, or certain other changes of form or of activity of the Aurora. But generally the effect of these phenomena upon the needle was not visible immediately; but in about half an hour or an hour the needle had obtained its maximum of deviation. From this its return to its former position was very gradual, seldom regaining it before the following morning, and frequently not until the afternoon, unless it was expedited by another arch of the Aurora operating in a direction different from the former one. The magnetic needle in the open air was disturbed by the Aurora whenever it approached the zenith. Its motion was not vibratory (as observed by Mr. Dalton), perhaps owing to the weight of the card. It moved slowly to the E. or W. of the magnetic meridian, and seldom recovered its original direction in less than eight or nine hours. The greatest extent of its aberration was 45′. The arches of the Aurora were remarked commonly to traverse the sky nearly at right angles to the magnetic meridian; but deviation was not rare, and it was considered that the different positions of the Aurora had considerable influence on the direction of the needle. When an arch was nearly at right angles to the magnetic meridian, the motion of the needle was towards the W. This motion was greater when the extremity of the arch approached from the west towards the magnetic north. A westerly motion also took place when the extremity of an arch was in the true north, or about 36° to the west of the magnetic north. The motion of the needle was towards the east when the same end of an arch originated to the southward of the magnetic west, and when of course its opposite extremity approached nearer to the magnetic north. In one case only a complete arch was formed in the magnetic meridian. In another the beam shot up from the magnetic north to the zenith. In both these cases the needle moved towards the west. The needle was most disturbed on February 13, 1821, at a time when an Aurora was distinctly seen passing between a stratum of clouds and the earth. Sometimes the needle deviated though no Aurora was visible; but it was uncertain whether there might not have been a concealed Aurora at the time. Clouds were sometimes observed during the day to assume the form of the Aurora, and deviations of the needle were occasionally remarked at such times. An Aurora sometimes approached the zenith without producing any change of position of the needle; while at other times a considerable alteration took place, though the beams or arches did not come near the zenith. The Aurora was frequently seen without producing a perceptible effect on the needle. At such times it was generally an arch or a horizontal stream of dense yellowish light with little or no internal motion. The disturbance of the needle was not always proportionate to the agitation of the Aurora, but was always greater when the quick motion and vivid light were observed to take place in a hazy atmosphere. In a few instances the needle commenced at the instant a beam started from the horizon upwards; and its return was according to circumstances. If an arch formed immediately afterwards, having its extremities placed on opposite sides of the magnetic north and south to the former one, the return of the needle was more speedy, and it generally went beyond the point from which it first started. When the disturbance was considerable, it seldom regained its usual position before 3 or 4 P.M. on the following day. On one occasion only the needle had a quick vibratory motion (between 343° 50′ and 344° 40′). The disturbance produced by the Aurora was so great that no accurate deductions as to diurnal variation could be made.

Magnetic observations on board the 'Tegetchoff.'

Payer, in his 'New Lands within the Arctic Circle' (vol. i. pp. 327, 328), gives the result of the magnetic observations on board the Austrian ship 'Tegetchoff' in the years 1872-74, made by means of a magnetic theodolite, a dipping-needle, and three variation instruments. The extraordinary disturbances of the needle rendered the determination of exact mean values for the magnetic constants impossible. The following were the principal results of these observations:—

Disturbances great.

(1) The magnetic disturbances were of extraordinary magnitude and frequency.

Greater as the rays were rapid. Quiescent arches exercised no influence.

(2) They were closely connected with the Aurora, and they were greater as the motion of the rays was more rapid and fitful and the prismatic colours more intense. Quiescent and regular arches, without changing rays or streamers, exercised mostly no influence on the needle.

Declination-needle, effects on.

(3) In all the disturbances the declination-needle moved towards the east, and the horizontal intensity decreased while the inclination increased.

Sir John Franklin sums up his information as to the needle to much the same effect, viz. that brilliant and active coruscations cause a deflection almost invariably if they appear through a hazy atmosphere and if the prismatic colours are exhibited in the beams or arches. On the contrary, when the air is clear and the Aurora presents a steady dense light of a yellow colour and without motion, the needle is often unaffected by its appearance.

Parry's experience.

Parry (Third Voyage) found his variation-needle (extremely light and delicately suspended) in no instance affected by the Auroræ; but he seems to have principally met with the quiescent form of that phenomenon.

M. Lottin, the French savant (whose description of an Auroral display has been given in Chapter II.), observed in the North Sea, between September 1838 and April 1839, while the sun was below the horizon, 150 Auroræ. During this period 64 were visible, "besides many which a cloudy sky concealed, but the presence of which was indicated by the disturbances they produced upon the magnetic needle" (Lardner's 'Museum of Science and Art,' vol. x. p. 189).

Grand displays accompanied by motion of needle to the west.

It has been remarked by some observers that grand displays of the Aurora are frequently preceded or accompanied by an extraordinary motion of the needle to the westward.

Captain Maguire found at Point Barrow (1852-54) that the appearance of the Aurora in the south was connected with the motion of the magnet to the east of the magnetic north, and if in the north to the west of the same.

Solar disturbances and Aurora.

On an occasion in 1859 great solar disturbances were observed, the Greenwich magnets were much disturbed, and a fine Aurora was visible.

Cipoletti's observation.

Cipoletti, of Florence, remarks on the strong magnetic disturbances at Vienna and Munich during the Auroræ of 4th February, 1872, and 4th February, 1874.

Dr. Thompson concludes that cylinders of Aurora cannot be doubted to be magnets.

Dr. Thompson, in his 'Annals of Philosophy,' vol. iv. p. 431 (1814), mentions as an authenticated fact that during the prevalence of the Aurora the magnetic needle was frequently observed to become unsteady, and (p. 432) concludes that cylinders of Aurora cannot be doubted to be magnets. The only three bodies capable of assuming magnetic properties are iron, nickel, and cobalt. When meteors are considered, it is not altogether extravagant to conjecture that bodies similar in their nature to some of the solid bodies which constitute our globe may exist in some unknown state in the atmosphere.

During the Aurora of 13th May, 1869, the declination at Greenwich varied 1° 25', while the vertical force experienced four successive maxima, and the greatest oscillation amounted to 0·04 of the total mean value. The horizontal force varied only 0·014 of its mean value.

During the Aurora of 15th April, 1869, the declination at Stonyhurst varied 2° 23' 14" in nine minutes.

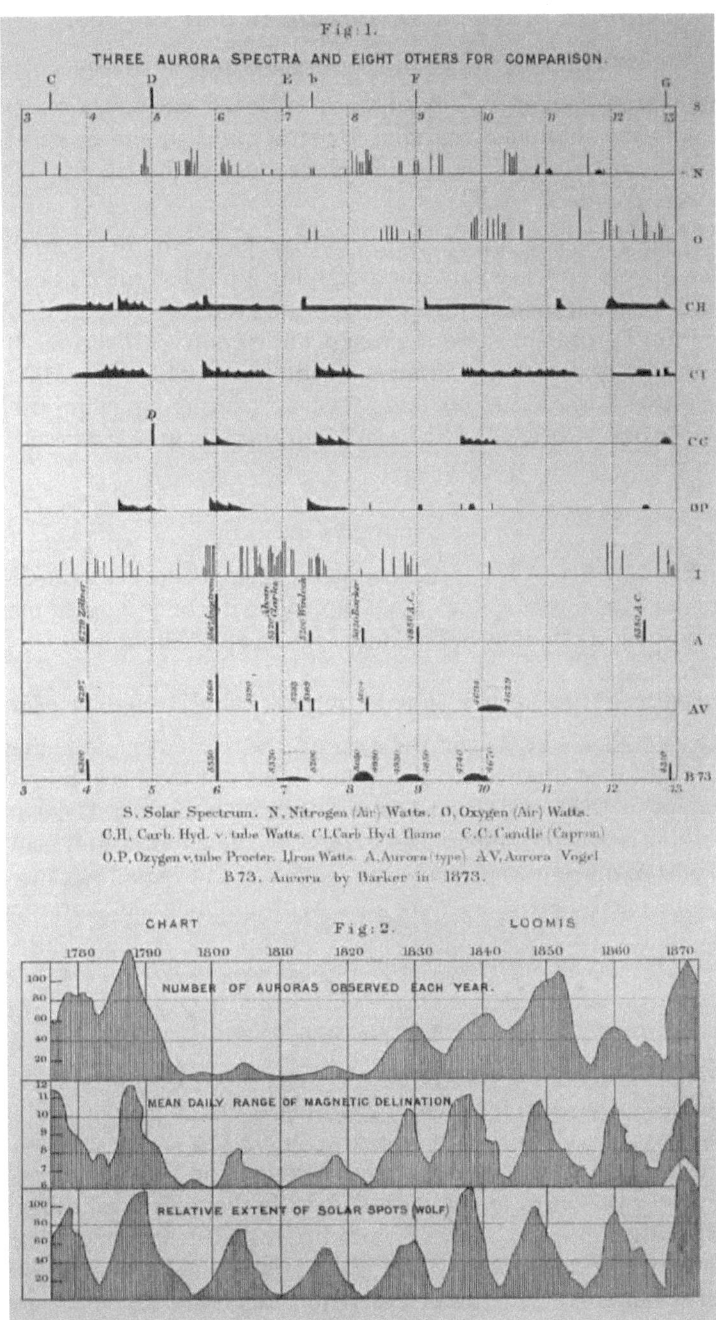

Plate IX.

COMPARED AURORA AND OTHER SPECTRA. LOOMIS'S CURVES OF AURORAS, MAGNETIC DECLINATION, AND SOLAR SPOTS

Auroræ, Magnetic Disturbances, and Sun-spots.

Auroræ, magnetic disturbances, and sun-spots in Italy.

Auroræ were frequent in Italy in April 1871. On the 10th a remarkable one was seen, with declinometer deflected towards the east, and 63 sun-spots were counted. On the morning of the 10th the deflection continued, and at midday 97 sun-spots were counted.

On the 18th a brilliant Aurora lasted to 10 o'clock at night. From this time till the 23rd the Aurora appeared constantly, giving a reddish tinge in the north and north-west. A brilliant display took place on the evening of the 23rd. On the evenings when the Aurora appeared the magnetometers were disturbed throughout Italy, and ended by a violent agitation during the whole of the 24th. Sun-spots were observed at Rome, Palermo, and Moncalieri, but the greater number on the days of the Auroræ. A brilliant display at Moncalieri on June 18 was accompanied by very violent magnetic disturbance.

Proctor's sun-spots and Aurora.

September 25, 1870, Mr. Proctor counted 102 spots on the solar disk; and on the night of the 24th and morning of the 25th an Aurora of unwonted magnificence was visible at various stations in England, France, and Germany.

Sun-spots and the magnet. 11 years' period. Schwabe's sun-spot period.

With respect to sun-spots and the magnet, the frequency of magnetic storms, causing oscillation of the needle, gradually increased from a minimum in 1843 to a maximum in 1848, giving a variation of something near 11 years altogether. Schwabe observed the sun-spots for 24 years, and found they had a regular maximum and minimum every five years, and that the years 1843 and 1848 were minimum and maximum years coinciding with the magnetic variation at those periods.

Prof. Loomis considers connexion established between magnetic declination, auroral displays, and sun-spots.

Professor Loomis ('American Journal of Science,' vol. v. April 1873) considers that a comparison between the mean daily range of the magnetic declination and the number of Auroras observed in each year, and also with the extent of the black spots on the surface of the sun, establishes a connexion between these phenomena, and indicates that auroral displays (at least in the middle latitudes of Europe and America) are subject to a law of periodicity, that their grandest displays are repeated at intervals of about 60 years, and that there are also other fluctuations, less distinctly marked, which succeed each other at an average interval of about 10 or 11 years, the times of maxima corresponding quite remarkably with the maxima of solar spots.

Illustrative table of coincidences.

An illustration of the result of these observations is given on Plate IX. fig. 2.

The curves are in close correspondence, and the coincidence at the times of maximum and minimum is remarkable. The auroral maximum generally occurs a little later than the magnetic maximum; and the connexion between the auroral and magnetic curves appears somewhat more intimate than between the auroral and sun-spot curves.

Prof. Loomis considers a sun-spot a solar disturbance affecting the earth's magnetism.

Professor Loomis contends "that the black spot is a result of a disturbance of the sun's surface, which is accompanied by an emanation of some influence from the sun, which is almost instantly felt upon the earth in an unusual disturbance of the earth's magnetism, and a flow of electricity, developing the auroral light in the upper regions of the earth's atmosphere."

Carrington and Hodgson's observations of bright spots on the sun, accompanied by magnetic disturbance at Kew, and followed by wide-spread Auroræ.

This connexion between the sun's spots and the earth's magnetism has been considered as proved; and one instance at least of an intense disturbance and outbreak of the sun's surface having been observed simultaneously with the occurrence of a terrestrial magnetic storm is a matter of record. This will be found detailed in the 'Monthly Notices of the Royal Astronomical Society,' vol. xx. pp. 13 and 15, and is so interesting in its character that it may be briefly referred to here. Mr. R. C. Carrington, September 1, 1859, 11h 18m, while observing and drawing a group of solar spots, saw suddenly two patches of intense bright light break out in the middle of the group. The brilliancy was fully equal to that of direct sunlight. Seeing the outbreak was on the increase, Mr. Carrington left the telescope, to call some one to witness it. On his return within sixty seconds it was nearly concluded. The spots travelled from their first position, and vanished as two rapidly fading dots of white light. In five minutes the two spots traversed a space of about 35,000 miles. Mr. Carrington found no change in the group itself. His impression was that the phenomena took place at an elevation considerably above the general surface of the sun, and above and over the great group of spots on which it was seen projected. It broke out at 11h 18m, and vanished at 11h 23m. Mr. R. Hodgson independently on the same day, and at close upon the same time, saw a very brilliant star of light, much brighter than the sun's surface, most dazzling to the protected eye, illuminating the upper edges of the adjacent spots and streaks. The rays extended in all directions, and the centre might be compared to α Lyræ when seen in a large telescope. It lasted for some five minutes.

At the very moment of this solar disturbance the instruments at Kew indicated a *magnetic storm*; and Proctor, in his volume on the Sun, page 206, details how this magnetic storm was accompanied by very widely-spread indications of electrical disturbance in many parts of the globe. Vivid Auroræ were seen not only in both hemispheres, but in latitudes and places where they are seldom witnessed. Rome, Cuba, and the West Indies, the tropics within 18° of the equator, and even South America and Australia, are thus referred to for displays. At Melbourne, on the night of September 2nd, the greatest Aurora ever seen there made its appearance.

It was observed, too, that magnetic communication was at the same time disturbed all over the earth. Strong currents, continually changing their direction, swept along the telegraphic wires. At Washington and Philadelphia the signal-clerks received severe shocks, and the wires had to give up work. At a station in Norway the transmitting apparatus was set fire to; and at Boston, in North America, a flame of fire followed the pen of Baine's electric telegraph.

Mr. John Allan Broun's magnetic oscillation-curves; showing that the sun's magnetic action has lately become more constant. In diagram, curves gradually flatten.

In an interesting communication to 'Nature' (January 3rd, 1878), entitled "The Sun's Magnetic Action at the Present Time," Mr. John Allan Broun has contributed some magnetic oscillation-curves, deduced from observations made in the Trevandrum Observatory (nearly on the magnetic equator), by which, if confirmed by other observations, it would appear that the sun's magnetic action has lately become gradually more constant. The curves are three in number,—no. 1 for the years 1855-58, no. 2 for the years 1865-68, no. 3 for the years 1874-77. In no. 1 curve the minimum is very clearly marked by two points corresponding to April 1 and May 1, 1856, and there is little difference in the rapidity with which the curve descends to and ascends from the minimum. In no. 2 curve the epoch of minimum is by no means so well marked; it occurs between the points for April 1 and September 1, 1866. There is also a considerable difference in the rapidity of variation in the descending and ascending branches of the curve. The descent is nearly as rapid as in curve no. 1; but the ascent is very much slower. In curve no. 3 the lowest point is that for December 1, 1875; but it is even now, with points a year and a half later, difficult to say whether this is the minimum or not, the point for January 1, 1877, being only 0·02 (two hundredths of a minute of arc) higher. In this curve the change of range in diurnal oscillation is quite insignificant from November 1, 1874, to April 1, 1877, an interval of three years and five months. In the diagram given by Mr. Broun the curves show themselves gradually flattening, no. 3 being almost a straight line.

Mr. Broun never found an Aurora without a corresponding irregularity in the declination-needle.

Mr. Broun remarks upon the report of Sir George Nares as to the insignificant nature of the Auroræ seen in the Arctic Expedition in the winter of 1875-76, and the accompanying statement that, as far could be discovered, they were totally unconnected with any magnetic or electric disturbance; and states, as the result of his own experience in the south of Scotland, that several of the Auroræ observed by him were of the very faintest kind, "were traces" which he could never have remarked had he not been warned by very slight magnetic irregularities to examine the sky with the greatest attention. Again, in no case had he seen the faintest trace of an Aurora without finding at the same time a corresponding irregularity in the movement of the force or declination-magnet.

Prof. Piazzi Smyth comments on variance in the cycles.

Prof. Piazzi Smyth, commenting on this article, makes the inquiry how the sun-spot cycle and the terrestrial magnetic oscillation cycle can be considered as agreeing, the sun-spot cycle, according to Prof. Wolf, being 11·111 years, and the magnetic cycle 10·5 years according to Mr. Broun.

M. Faye's remarks to a similar effect.

Another correspondent writes and quotes M. Faye, in 'La Météorologie Cosmique,' for the remark, "La période des taches portée à 11 ans ·1 par M. Wolf n'étant pas égale à celle des variations magnétiques (10 ans ·45), ces deux phénomènes n'ont aucun rapport entre eux."

Mr. Broun's rejoinder and explanation.

Mr. Broun, in a further letter, rejoins that if we could accept Dr. Wolf's view we should find that the mean duration of a cycle for *both* phenomena since 1787 would be 11·94 years, while the sun-spot results for eight cycles determined by Dr. Wolf during eighty years before 1787 give 10·23 or, if we take nine cycles, 10·43 years for the mean duration. It is by mixing these two very different means that the Zurich philosopher finds 11·1 years, a mean which Mr. Broun considers can evidently have no weight given to it. On the other hand, if Dr. Wolf is in error (as Mr. Broun believes he is) as to the existence of a maximum in 1797, the mean durations for the eighty years after and for the eighty years before 1787 agree as nearly as the accuracy of the determinations for the beginning of the eighteenth century will permit. Mr. Broun then repeats his conviction that the sun-spot maxima and minima are really synchronous with those of the magnetic diurnal observations.

Mr. Jenkins's explanation of Prof. Loomis's chart.

Mr. B. G. Jenkins, in a letter to 'Nature,' refers Prof. Smyth to Prof. Loomis's chart of magnetic oscillations given in Prof. Balfour Stewart's paper in 'Nature' (vol. xvi. p. 10), for the purpose of showing that there are exactly seven minimum periods from 1787 to 1871, the mean of which is twelve years, the mean of the seven corresponding maximum periods being 11·8 years. The true magnetic declination-period is, then, the mean of these, viz. 11·9 years. In exactly the same manner he finds that the mean period of sun-spots is 11·9 years.

Jupiter's suspected connexion with sun-spots.

The auroral displays also have the same period. Mr. Jenkins also refers to Wolf, De La Rue, Stewart, and Loewy, as having stated their belief that Jupiter is the chief cause in the production of sun-spots, and draws attention to the period of 11·9 years as being Jupiter's anomalistic year, or the time which elapses between two perihelion passages.

Infrequency of Auroræ and absence of sun-spots in 1876-78.

The infrequency of Auroræ during the years 1876-78, and a corresponding comparative absence of sun-spots, may be added to the evidence on the subject. I have seen no account of important Auroræ during the years mentioned, and day after day has recently (1878) passed with a perfectly clean sun-disk.

Aurora and Electricity.

Aurora and electricity. Sir John Franklin's experience with electrometer.

Sir John Franklin failed to get indications of electricity connected with the Aurora with a pith-ball electrometer; but with another form of electrometer specially constructed for the purpose he seems to have got some, though not very strong or regular, indications of repulsion between the needle of the instrument and the conductor when Auroræ were seen. He does not decide whether the electricity was received from or summoned into action by the Aurora.

Parry's experience.

Parry, at Fort Bowen, with a gold leaf electroscope connected with a chain attached by glass rods to the skysail mast-head, 115 feet above sea-level, found no effect.

Dr. Allnatt's experience, February 4, 1872.

Dr. Allnatt, at Frant, during the display of 4th February, 1872, found the earth's electricity so powerful that the gold leaves of the electrometer remained divergent for a considerable time.

M'Clintock found electroscope affected in Baffin's Bay and Port Kennedy.

M'Clintock observes that on six occasions of Aurora in Baffin's Bay, the electroscope was strongly affected, and on three occasions of Aurora at Port Kennedy. The electricity was always positive.

Dec. 18.—Dr. Walker called him to see the electroscope. The charge was at first weak, but afterwards strong enough to keep the leaves diverged. Dr. Walker found two periods of minimum electrical disturbance about 9 P.M. and noon.

Electric currents in telegraphic wires during Auroræ.

Electric currents have been reported as produced in telegraph wires during Auroræ. Though transient they are said to be often very powerful, and to interrupt the ordinary signals. Loomis (Sillim. Journ. vol. xxxii.) mentions cases where wires have been ignited, brilliant flashes produced, and combustible materials kindled by their discharge.

Here, too, we may note the account of electric phenomena in the case of the Aurora Australis described (*antè*, p. 28) by Mr. Proctor.

Mr. George Draper's report as to disturbed condition of the Indian Submarine and other cables during Aurora of February 4, 1872.

Mr. George Draper, of the British-Indian Submarine Telegraph Company, speaking of the Aurora of February 4, 1872 (and writing to the 'Times' under date February 5th), states that the Aurora visible in London was also visible at Bombay, Suez, and Malta, and that the Company's electrician at Suez reported that the earth-currents there were equal to 170 cells (Daniell's battery), and that sparks came from the cable. The electrical disturbances lasted until midnight, and interrupted the working of both sections of the British Indian Cable between Suez and Aden, and Aden and Bombay. For some days previously the signals on the British Indian cables had been much interfered with by electrical and atmospheric disturbances.

At Malta there was a severe storm on the morning of the 4th, so that it was necessary to join the cable to earth for some hours, and the Aurora was very large and brilliant there.

The electrical disturbances on the cables in the Mediterranean and on those between Lisbon and Gibraltar, and Gibraltar and the Guadiana, were also very great. The signals on the land line between London and the Land's End were interrupted for several hours on the night of the 4th by atmospheric currents. Similar effects accompanied the displays of Oct. 24 and 25, 1870.

Aurora and Meteoric Dust.

Aurora and meteoric dust. Theories of Dr. Zeyfuss and M. Gröneman.

A theory has been propounded independently by Dr. Zeyfuss and by M. Gröneman, of Gröningen, according to which the light of the Aurora is caused by clouds of ferruginous meteoric dust ignited by friction with the atmosphere. Gröneman shows that these might be arranged along the magnetic curves by the action of the earth's magnetic force during their descent, and that their influence might produce the observed magnetic disturbances.

Ferruginous dust in the Polar Regions.

The arches might be accounted for by the effects of perspective; and the iron spectrum shows correspondence with some of the lines of the Aurora. Ferruginous particles have been found in the dust of the Polar regions according to Professor Nordenskiöld, but whether derived from stellar space or from volcanic eruption is uncertain. A difficulty has been suggested that while meteors are more frequent in the morning, or on the face of the earth which is directed forward on its orbit, the reverse prevails in the case of Auroræ. Gröneman meets this by supposing that in the first case the velocity may be too great to allow of arrangement by the earth's magnetic force. He accounts for the infrequency of the Aurora in equatorial regions by the weakness of the earth's magnetic force, and the fact that when it does occur the columns must be parallel to the earth's surface.

Baumhauer's proposition.

Baumhauer (Compt. Rend. vol. lxxiv. p. 678) advances, as regards Polar Auroræ, the proposition, that not only solid masses large and small, but clouds of "uncondensed" (meteoric) matter probably enter our atmosphere.

If from our knowledge of the meteoric stones which fall to the earth's surface we may draw any conclusion respecting the chemical constitution of these clouds of matter, it would appear that they may contain a considerable portion of the magnetic minerals iron and nickel. Let such a cloud approach our earth, regarded as a great magnet, it would be attracted towards the Pole, and, penetrating our atmosphere, the particles which have not been oxidized, and are in a state of extremely fine division, would by their oxidation generate light and heat, producing the polar Auroræ. Baumhauer suggests it would be interesting in support of this theory to detect in the soil of polar areas the presence of nickel. The presence of iron and nickel in meteoric masses in considerable quantities is frequent; and cases are also on record by Eversmann of hailstones containing crystals of a compound of iron and sulphur, by Pictet of hailstones containing nuclei which proved to be iron, and by Cozari of hailstones containing nuclei of an ashy-grey colour, the larger ones of which were attracted by the magnet, and found to contain iron and nickel. Nickel was found by Reichenbach in parts of Austria on hills consisting of beds of sandstone and limestone, and quite free from metallic veins.

Mr. Lefroy's description of a phenomenon ascribed by him to streams of cosmic dust.

Mr. J. W. N. Lefroy, in 'Nature,' describes a phenomenon seen by him at Fremantle, West Australia, in the month of May, which he designates "A Lunar Rainbow, or an Intra-lunar convergence of Streams of slightly illuminated Cosmic Dust?"

It lasted about three quarters of an hour, and consisted of one grand central feather, of very bright white cloud, springing out of the horizon at W.N.W., and crossing the meridian at about 20° north of the zenith, with a width of 7° to 8°.

On either side of this was a system of seven or eight minor beams of light, extending from the W. to the E. horizon, subtending a chord common to themselves and to the main stream, and converging towards the axis of the central stream so as to intersect it at a point about 30° or 40° below the western horizon, at which the whole system subtended an azimuth of about 20°. Near the zenith, where its transverse section was a maximum, that section subtended an angle of about 40°.

The idea strongly suggested itself to Mr. Lefroy of converging streams of infinitely minute particles of matter passing through space at a distance from the earth at which its aerial envelope may have still a density sufficient by its resistance to give cosmic dust passing through it that illumination which it possessed. In about twenty minutes the streams of light had attained their maximum brightness. Their apparent figure was that of a nearly circular (slightly flattened) arc of an amplitude of 15° or 20°, as viewed from the middle point of its chord.

The brightness and the convergence of the streams were both more marked towards the western horizon than the eastern. This same phenomenon was described in the 'South-Australian Register' as a beautiful lunar rainbow visible in the western heavens.

Mr. Lefroy and other observers concurred in the impression that the minor lateral streams on the N. side of the main one intervened between the earth and the moon, and that one or more of them in their slow vibrations swept the surface of the moon and sensibly obscured its light. There can be hardly any question that the phenomenon observed was in fact an Aurora.

Suggestion as to collecting iron and nickel particles from the atmosphere.

It may be a question whether iron and nickel particles of meteoric origin do not ordinarily exist in the atmosphere in a greater degree than we suspect, and might be detected if special means, such as magnets, plates of glass covered with glycerine, &c., were adopted for the purpose of collecting and examining the cosmic dust. Larger gatherings than usual of iron and nickel particles during the presence of Auroræ would be in support of Mr. Lefroy's theory.

The Aurora and the Planets Venus and Jupiter.

The Aurora and planets Venus and Jupiter. The planet Venus's halo during Aurora.

During a brilliant Aurora seen at Sunderland, February 8, 1817 ('Annals of Philosophy,' p. 250), about 8 o'clock, Venus was about 8° above the horizon, and displayed a very peculiar appearance. Her rays passed through a thin mist or cloud, probably electric, of a deep yellow tint. Her apparent magnitude seemed increased, and a halo was formed round her as sometimes appears round the moon in moist weather; but the stars that were in that part of the heavens shone with their accustomed brilliancy.

Dr. Miles's observation of Venus during an Aurora.

The Rev. T. W. Webb, in his 'Celestial Objects' (1859), p. 43, quoting from the Philosophical Transactions, mentions that, "January 23rd, 1749-50, there was a splendid Aurora Borealis about 6 P.M. The Rev. Dr. Miles, at Tooting, had been showing Jupiter and Venus to some friends with one of Short's reflectors, greatest power 200, when a small red cloud of the Aurora appeared, rising up from the S.W. (as one of a deeper red had done before), which proceeded in a line with the planets and soon surrounded both. Venus appearing still in full lustre, he viewed her again with the telescope without altering the focus, and saw her much more distinctly than ever he had done upon any occasion. His friends were of the same opinion. They all saw her spots plain (resembling those in the moon), which he had never seen before, and this while the cloud seemed to surround it as much as ever."

I think this effect might perhaps have arisen from the Aurora acting as a screen, and removing the glare with which so bright an object as Venus is always accompanied; but the case is a singular one, and one would be glad of further experience. I suggested observations on this head during Sir Geo. Nares's Arctic Expedition; but the suggestion, for some reason of which I am not aware, was not included in the official instructions issued.

Brightness of stars during Auroræ.

Remarks are frequent of the brightness of stars as seen through Auroræ. Payer, of the Austrian Expedition, remarks that falling stars passed through the Aurora without producing any perceptible effect or undergoing any change.

Aurora of Oct. 24, 1870, and Jupiter.

A grand display of the Aurora took place 24th October, 1870. About this time the belts of Jupiter were observed to be highly coloured. As observed by me on the night of November 2, 1870, at 9 P.M., with an 8¼-in. Browning reflector, achromatic eyepieces 144, 305, and 450, the equatorial zone was of a distinctly dark ochre colour, deepening to red-brown as it approached the lower (N.) edge. Two thin belts above were slate-purple, and a darker belt below was of a deep purple colour.

According to Lassell and others, Jupiter's belts exhibit the brightest colours at period of Auroræ.

Lassell, Proctor, and others have reported Jupiter's belts to exhibit the brightest colours at the period of Auroræ. Mr. Browning gives a drawing of Jupiter as seen on January 31, 1870 (a year noted for Auroræ), with the belts brightly coloured. The finest view of Jupiter I ever had was on the 8th February, 1872 (a fine Aurora was on the 4th), when, with the 8¼-inch Browning reflector, I saw the whole surface of the planet (by glimpses) cloud-mottled. The equatorial belt was, however, then slightly tinted only. In Dr. Miles's observation (p. 66) he does not seem to have noticed the colouring of Jupiter's belts.

Infrequency of Auroræ and lightness in tint of Jupiter's belts.

The three past years, 1876, 1877, and 1878, have been distinguished by the infrequency of Auroræ; and Jupiter's equatorial zone and belts have been mainly reported of light tints.

The subject apparently deserves more attention than it has hitherto received.

The Aurora and the Zodiacal Light.

The Aurora and the Zodiacal Light. Ångström's observation on spectrum.

Ångström in 1867 found the spectrum of the Zodiacal Light to be monochromatic, consisting of a single line in the green, to which he assigned approximately the position 1259 on Kirchhoff's scale, the same that he had determined for the green line of the Aurora Borealis; and Respighi, on the Red Sea, on the evening of the 11th and the morning of the 12th January 1872, perceived in the Zodiacal Light not only this green line, but near it, towards the blue, a band or zone of apparently continuous spectrum.

Respighi's at Campidoglio.

At the Observatory of the Royal University of Campidoglio, February 5th, 1872, Respighi, at 7 P.M., was able to discern the same spectrum; and on directing the spectroscope to other points he found that this spectrum showed itself in all parts of the heavens from the horizon to the zenith, more or less defined in different parts, but everywhere as bright as in the Zodiacal Light. The Observatory Assistant, Dr. di Legge, likewise observed this spectrum distinctly in various parts of the heavens. Respighi's observations corroborating Ångström's in 1867, appeared to him to demonstrate the identity of the Zodiacal Light with the Aurora, and to establish the identity of their origin.

Pringle thinks the Aurora may be considered as allied to the Zodiacal Light.

Pringle, in a letter to 'Nature' from South Canara, October 3, 1871, alludes to the Aurora as being considered by many allied to the Zodiacal Light, and does not think the evidence then hitherto adduced against the theory at all conclusive. He says:— "Assume the auroral light to consist of solid particles of matter, planet dust, shining by reflected light, and it is not difficult to imagine the Aurora playing amongst these tiny worlds, each of which would have its own small magnetic system swayed like our own by the monster magnet the sun."

Phosphorescence of sky when Zodiacal Light has been seen bright.

He notices he has never found it to have a decided outline, nor traced it east or west to 180° from the sun. He also refers to others having noticed that when the Zodiacal Light has been seen unusually bright, a "phosphorescence" of the sky was everywhere visible.

Pringle failed to find bright lines or bands in the Zodiacal Light.

He does not seem at that time to have examined the matter spectroscopically; and on June 23, 1872, he writes again, pointing out the peculiarity in Respighi's observation that the green line was seen everywhere as bright as in the Zodiacal Light, and suggesting that it was due to a concealed Aurora present at the time of Ångström's and Respighi's observations. He further states he had examined the Zodiacal Light with a Browning 5-prism spectroscope (I presume a compound direct-vision form is meant) since the last December, and, brilliant as the phenomenon had frequently been, failed to detect the slightest appearance of bright lines or bands. A faint diffuse spectrum about as intense as that of a bright portion of the Milky Way was all he had obtained.

Prof. Piazzi Smyth confirms this.

Professor Piazzi Smyth, in the clear sky of Italy, and with an instrument specially designed for showing faint spectra, found no lines or bands, but only a faint continuous spectrum extending from about midway between D and E in the solar spectrum to nearly F (see Plate V. fig. 3, in which the continuous spectrum is graphically shown, white on a black ground).

Colour of the Zodiacal Light.

It may here be mentioned that the Zodiacal Light is usually described as, in these latitudes, of a golden yellow or pale lemon tinge.

Rev. Mr. Webb's observation, February 2, 1862. He found no green line of the Aurora.

On one occasion, however, it has been described as not having this tinge, but rather resembling the light of the Milky Way, but brighter. On another occasion I saw the whole cone of a crimson hue without any mixture of yellow. The Rev.

Mr. Webb thought that a display seen at Hardwick Vicarage, February 2nd, 1862, showed a ruddy tinge not unlike the commencement of a crimson Aurora—"it was certainly redder or yellower than the galaxy." He examined it with a pocket spectroscope which would show distinctly the green line of the Aurora (probably Browning's miniature), but nothing of the kind was visible, nor could any thing be traced beyond a slight increase of general light, which, on closing the slit, was extinguished long before the auroral band would have become imperceptible.

A. W. Wright's observations and conclusions.

A. W. Wright examined the Zodiacal Light with a Duboscq single-prism spectroscope, the telescope and collimator having a clear aperture of 2·4 centimetres, magnifying-power of telescope 9 diameters. Special precautions were taken about the observations, and the conclusions arrived at were:—

(1) The spectrum of the Zodiacal Light is continuous, and is sensibly the same as that of faint sunlight or twilight.

(2) No bright line or band can be recognized as belonging to this spectrum.

(3) There is no evidence of any connexion between the Zodiacal Light and the Polar Aurora.

Polarization of Zodiacal Light. Burton's observation confirmed by Wright and Tacchini.

The Polarization of the Zodiacal Light has been already referred to under the head of "Polarization of the Aurora:" but it may be here noted that Mr. Burton's observation of polarization of the light there mentioned has been confirmed by Wright and Tacchini, and the presence of reflected sunlight established. In this respect it differs from the Aurora, in which no trace of polarization has hitherto been detected; and looking at this, and at the weight of evidence in the spectroscopic observations, the theory of a connexion between the Aurora and the Zodiacal Light must, as the matter stands, be given up.

CHAPTER VII.

AURORA-LIKE PATCHES ON THE PARTIAL-LY-ECLIPSED MOON.

Aurora-like patches on the partially-eclipsed moon, Feb. 27, 1877.

In anticipation of the total eclipse of the Moon on the 27th February, 1877, several articles appeared in the leading journals of the day describing, for the public benefit, the appearances which might be expected during the occurrence of the phenomenon.

Formerly it was thought the moon was illuminated by auroral light.

Among these was one by Mr. R. A. Proctor, in which the following passage occurs:—"That dull, or occasionally glowing red colour, shown by the moon when she is fully and even deeply immersed in the shadow of the earth, is a phenomenon whose explanation is not without interest. Formerly it was thought that the moon possessed an inherent light, or *perhaps was illuminated by auroral light*, which only became discernible at the time of total eclipse. Indeed even Sir W. Herschel fell into the mistake of supposing this the only available explanation, having miscalculated the efficiency of the true cause."

Author's notes of the eclipse. Colour-tints described. A crimson-scarlet tint reminded author of an auroral glow.

This passage was only pointed out to me by a friend after the eclipse had actually taken place, and I had sent him some notes of what I then saw. My notes on the occasion comprised, amongst others, the following remarks:—"The tints of colour also during partial eclipse, owing, no doubt, to the moon's considerable altitude, were singularly bright and well contrasted. Silver-grey, dusky copper-red, and the same tint clearer and brighter were ranged side by side with a lovely jewel effect. *We noticed also at times a crimson-scarlet tint, deeper and less mixed with yellow than the copper colour.* This last tint reminded me much of a *crimson glow common to the Aurora*, and which I also once distinctly remarked (of course in a weaker degree) in the zodiacal light" (*antè*, p. 68).

Eclipse, Aug. 23-24, 1877. Sky clear, but eclipsed moon misty and indistinct until total obscuration. Succession of colours.

On the occasion of the eclipse of August 23-24, 1877, we were favoured at

Guildown, in common with many other places, by a singularly clear sky during the progress of the moon's obscuration and subsequent clearing. In the early part of the evening, however, the moon, from some cause (possibly atmospheric vapour), seemed to have, as the earth's shadow advanced on its disk, an unexpectedly misty and indistinct appearance, which lasted up to and including total obscuration. Golden yellow, yellow copper, dull copper, ruddy copper, and dull red were successively the principal colours observed at different times and at various portions of the moon's surface.

As shadow passed off, indistinctness gave way to a sharpness of the moon's features as seen through shadow. Two patches of crimson light described.

After referring to some spectroscopic appearances, my notes then ran on thus:—"As the shadow began to pass off, and the bright sharp crescent of the illuminated portion of the moon to appear, the general aspect of the moon's disk seemed to me to greatly change. The certain amount of indistinctness noticeable during approach and continuance of totality, gave way to a considerable sharpness of the moon's features as seen through the shadow. The shadowed part glowed with a richer copper tint, on which were seen dark, almost black, spots and patches." Then follows a description of these; and the notes continue:—"Two features here struck me—the one a continuation of the upper limb of the illuminated crescent, so that it seemed to form a bead of light just on the centre of the upper edge of the moon; the other *two patches of crimson light,* similar to those I described as having been seen in the last total eclipse. One of these, quite a small one, was just under the elongated bead before described; the other, a much larger and more diffused one, was seen towards the south-west limb of the moon, about midway between it and the centre. The spots or patches were of a decidedly crimson-red, in contrast to the ordinary copper-red of the disk, and were noticed by my friend as well as by myself."

Patches well seen in field-glass; lost in small refractor. They gradually deepened in tint.

These were eye observations. The patches were quite well seen (but not so brightly as with the eye) with a double achromatic field-glass. With a 3¼-inch Cooke refractor and low power, they seemed lost in the general moon tint; but they were then diminishing in brightness. From a comparison of my two sketches, the patches seem to have gradually deepened in tint, and we considered them to have disappeared in a like gradual manner.

Two sketches taken.

My first sketch was taken shortly after end of total phase; the second about ten minutes later. I have reproduced the original sketches in preference to any drawing prepared from them (Plate IV. figs. 2 and 3).

The patches did not last long, but were lost as the shadow swept off the moon. I saw nothing of the sort during the approach of or pending totality, nor until a small crescent of the moon began to appear behind the shadow.

I have looked for other accounts of these patches, but cannot find any. Most observers have described the deeper colour of the shadowed moon by the word "copper." Some extend this colour to red; but there is probably much in the state of the atmosphere affecting this.

Dec. 3, 1703, moon's colour described.

At Avignon, December 3, 1703, the moon appeared, pending eclipse, "extraordinarily illuminated and of a very bright red," while other and different features were seen at Montpelier.

On March 19, 1848, observers in England, Ireland, and Belgium described the moon's disk as "intensely bright coppery red." On the occasion of August 23-24, 1877, before mentioned, an article in one of the public papers described the moon's disk, during totality, as of a "dull copper colour."

Mr. Keye's observation.

Mr. Henry Keye, in the Engadine, at a height of 4500 feet above sea-level, and with the purest air, saw the partially covered moon (before totality) as a "dull copper colour."

Prof. Pritchard's. M. Faye's. Dr. Allnatt's at Frant.

Prof. Pritchard, writing from the Oxford University Observatory, says that at 12h 10m (about the time my sketches were taken) there was a good deal of light on the moon's following limb, and the colour was "more red than copper," and apparently redder than it had been at a similar distance of time before totality. Mons. Faye reported to the French Academy of Sciences that "a striking phenomenon not previously noticed was that the reddish tinge, resembling that of a fine sunset, was deepest at the margin of the disk, a circumstance which he could not explain." Dr. Allnatt, writing from Frant, says:—"At totality the moon's disk presented a most extraordinary appearance: the western limb was comparatively transparent, but the main body appeared as though enveloped in a semi-opaque clot of coagulated blood, through which the lunar features were dimly visible."

Observations as to the patches.

The observations of Prof. Pritchard and Mons. Faye point more immediately to redness; and this is the nearest approach I can find to the patches I noticed. These patches do not seem to me easy of explanation. They could not well be colours or details due to the actual surface of the moon itself. The moon, we are aware, has only a certain portion of the visible disk slightly tinted. The Mare Serenitatis is certainly of a slight green tinge; and to the Palus Somni and certain other districts is attributed a pale red or pink; but these tints could hardly have sufficed to produce the effect seen, as the patches were conspicuous for a bright and decided colour. The positions, moreover, did not correspond; while the ease with which other details of the surface were seen at the time would, if the tints had arisen from the surface itself, probably have enabled the circumstance to be detected.

Refraction of sun's rays not a satisfactory explanation.

The refraction of the sun's rays by passage through the earth's atmosphere is, too, not a satisfactory explanation. This, as judged by the appearance of the covered moon immediately before and at totality, gives a disk of shadow deeper in tone in the centre and lightening towards the edges, but in other respects fairly uniform, so that the whole disk seems to partake of the same tint and its graduations; and this is what might have been expected under the circumstances. The patches, on the other hand, were quite local.

Question of lunar atmosphere.

The theory of the moon's possessing no atmosphere whatever is now very generally, but perhaps too readily, received (mainly upon the evidence of the spectroscopic observations of occulted stars[9]), as there still seems a reasonable doubt whether our satellite may not possess an atmosphere, possibly rarefied, but yet sufficiently dense to permit of the formation of cloud or vapour.

Instance of patch of vapour or cloud on moon's surface.

A curious case, in which a patch of vapour or cloud was supposed to be detected on the moon's surface, is reported by the Rev. J. B. Emmett in a communication to the 'Annals of Philosophy' (New Series, vol. xii. p. 81). It is dated "Great Ouseburn, near Boroughbridge, July 5, 1826," the observation being made with "the greatest care with a very fine telescope."

On the 12th April 8h, while observing the part of the moon called Palus Mœotis by Nevelius, with an excellent Newtonian reflector of 6 inches aperture, at a particular part of the Palus, which he minutely describes, he saw, with powers 70 and 130, "a very conspicuous spot wholly enveloped in black nebulous matter, which, as if carried forward by a current of air, extended itself in an easterly direction, inclining a little towards the south, rather beyond the margin of Mœotis." April 13th 8h to 9h, the cloudy appearance was reduced both in extent and intensity, and the spot from which it seemed to issue had become more distinctly visible. On April 17th scarcely a trace of the nebulous matter remained; but so long after as June 10th 8h "a little blackness" remained about the spot. Mr. Emmett suggested "smoke of a volcano or cloudy matter." A copy of the drawing annexed to the paper is given on Plate X. fig. 10 (black patch on moon). If this observation was (as it certainly appears to be) critical and exact, there must have been a disturbance of the moon's surface, indicating some sort of cloud- or vapour-supporting atmosphere; and probably, for the purposes of Auroræ, an atmosphere of a very rarefied condition would suffice[10].

---

[9] The proof from occulted stars merely goes to the fact that the moon possesses no atmosphere *appreciable in that way*. It may still be a question whether there does not exist something of the kind, lying low and close to the surface, and possibly of a rarefied character, which would scarcely make itself visible by its effects in occultations. Cloud-vapour might form in an atmosphere of inconsiderable density.

[10] This observation is not without a certain amount of confirmation by more recent ones, in which certain lunar objects and regions have been suspected of mist or vapour. Mr.

Prof. Alexander's evidence in favour of a lunar atmosphere.

According to the 'New York Tribune,' at a recent semi-annual meeting of the American Academy of Sciences, Professor Alexander "brought forward a variety of evidence tending to indicate some envelope like an atmosphere for the moon. The evidence was principally drawn from observations during eclipses. The explanations usually offered for the bright band seen around the moon at such times was fully considered, and shown to be inadequate, though good as far as they would apply. The ruddy band of light is much too broad to be the sun's chromosphere. It was most apparent in those instances where the moon was nearest the earth. It would best be accounted for by supposing an atmosphere to the moon, a thin remnant of ancient nebulosity, comparable to that which accompanies the earth and gives rise to the appearance of the Aurora Borealis." Is it not, however, possible that the appearance might have arisen from Auroræ in action within the region of the earth's own atmosphere during the passage of the sun's rays through it at the time of the eclipse? The whole subject is difficult of explanation, and should be one of the points for attention on the occasion of the next total Lunar eclipse. It seemed to me appropriate for introduction into the present history of the Aurora, whatever its solution may ultimately be.

Mars and Jupiter.

In the case of Mars and Jupiter, whose atmospheres are sufficiently recognized, red- and scarlet-tinted patches are frequently noticed. In Mars this is generally attributed to the geological character of the surface of the planet itself; but I have observed on Mars's surface during the recent opposition a local rosy tint of a more diffused and indefinite character; and in the case of Jupiter the appearances seem almost always connected with the clouds' belts, as distinguished from the regions lying nearer to the planet's surface.

Prof. Dorna's "Lunar Aurora."

Professor Dorna, of Turin, ascribed a flickering light seen on the reddened disk of the moon during the Lunar eclipse of February 1877 to the action of a *Lunar Aurora*, holding that the refraction of the sun's rays within the cone of the earth's shadow was not an adequate explanation ('L'Opinione Nazionale,' March 3, 1877).

Spectroscopic observations bearing on the subject. Mr. Christie's observations at Greenwich.

Birt ('English Mechanic,' vol. xxviii. no. 725) mentions two—the cloud-like appearance of the white patch west of Picard, and the interior of Tycho, which at one time always misty and ill-defined, is now become perfectly distinct and sharply defined.

December 4, 1878, 4h 45m. I observed Klein's crater as a dull dark spot, larger than the true object; and while definition was good and other objects were well defined, "the floor of Klein's object, the oval spot near, and also Agrippa (especially), all had *an odd misty look as if vapour were in or about them*" ('English Mechanic,' vol. xxviii. no. 727). The mystery of different observers seeing and not seeing Klein's object on the same night is hardly to be accounted for by the angle of illumination.

The spectroscope might have afforded some information on the question; but my own telescopes (8¼ and 3¼ in.) were not of sufficient aperture to give a sensible spectrum of a portion of the moon's eclipsed surface, and my observations were chiefly made on the entire disk with hand-spectroscopes without a slit. Mr. Christie, at the Royal Observatory, Greenwich, made a set of observations during totality, and also during subsequent partial phase, with a single-prism spectroscope. During totality a strong absorption band was seen in the yellow, and the red and blue ends of the spectrum were completely cut off, while the orange was greatly reduced in intensity. The yellow and green were comparatively bright, and seemed to constitute the whole visible spectrum. The absorption band became narrowed as the end of the total phase approached, and during partial phase was reduced to a mere line. The red end of the spectrum was cut off by a dark band commencing about halfway from D to C, in which a black line was suspected. The bands observed were characteristic of the spectrum of light which has passed through a thick stratum of air. In the description of the spectrum of the Aurora in Part II., it will be seen that the conspicuous red and green lines of the Aurora are either coincident with, or very close to, some of these atmospheric lines. It does not appear that Mr. Christie examined the crimson patches specifically, nor that he saw bright lines on any part of the moon's eclipsed disk.

Mr. Pratt's notes of Lunar Eclipse, August 23, 1877.

Mr. Henry Pratt has also kindly handed me for use his notes of the Lunar eclipse of August 23, 1877, as seen at Brighton on a splendid night. They were made as the phenomenon progressed, are 58 in number, and in many instances only a few minutes, or even seconds, apart. A selection of them is here given:—9h 13m 50s, first contact of shadow. 9h 30m, shadow very dark; no details of disk easily seen. 9h 40m, first appearance of red. 9h 50m, *red* all over disk, except margin bluish and S. part green tint. 10h 2m, *a sudden brightening of whole disk*, in strong contrast to two minutes previously. 10h 15m, *E. limb much darker.* 10h 35m, *south pole decidedly brightest.* 10h 44m, *S.E. limb much brighter.* 10h 48m, *whole disk much darker.* 10h 51m, *S.E. limb brightening again.* 11h 1m, *N.E. limb brightening.* 11h 3m, *N.E. limb has darkened and brightened three times during last two minutes.* 11h 20m, N. pole has *darkened.* 11h 21m, N. pole has *brightened.* 11h 24m 30s, N. pole darker *red.* 11h 35m, N. pole *bright.* 11h 35m 30s, same *dark* and *red.* 11h 42m, N.E. limb especially bright for a few seconds, and then *reddened* and shaded again. 11h 49m, *S. pole reddened.* 12h 1m, *S.W. limb reddest part; S. pole red; N. pole paler red.* 12h 3m 50s, first appearance of E. limb (my first sketch was made shortly after this, and my second about ten minutes later). 12h 21m, a bright patch on N.N.W. separated from N. pole. 12h 24m, *S.W. region is reddest part of eclipse.* 12h 40m, *redness* of shadow fading out.

With a small Browning star-spectroscope Mr. Pratt saw the red and blue ends of the spectrum cut off, but nothing else. Mr. Pratt adds that the *red* colour was not an effect of contrast or an optical delusion in any way, as was proved by using at times a limited field containing only the red portion under examination. In reference to the curious brightening and darkening of the disk, and the change from time to time of local colour, he says that with much experience he has seen nothing of the same marked character on other occasions, and that "the whole matter was at the time astonishing to me, but none the less real." The local red patches seen by me seem also to have been observed by Mr. Pratt.

Mr. Pratt's observation on the floor of Plato.

As an addition to the instances of Tycho, Picard, &c., mentioned in the note on p. 73, Mr. Pratt has also sent me his notes of some observations by him, of "local obscuration of the floor of Plato." As somewhat condensed, they are as follows:—1872, July 16. While in other parts of the floor spots and streaks were well visible, "the N.W. portion was in such a hazy condition that nothing could be defined upon it." 1873, Nov. 1. 27 light streaks seen (7 new): the brightness of the streaks was in excess of their usual character, as compared with the craterlets; "an *obliteration* or *invisibility* of *all* the light streaks in the neighbourhood of craterlet no. 1 was very noticeable;" and also "a similar obliteration of the N. end of the streak called the Sector, near craterlet 3." 1874, January 1. 18 light streaks seen, including 3 new, "some of which outshone other longer known ones. This was curious; for had they been as bright within the last two years as on this occasion I must have noticed them." Mr. Pratt points out, as worthy of remark, that some months previous to November 1st, 1873, neither craterlets nor streaks on the floor of Plato "had maintained their previous characteristic brightness,"—a fact which he thinks ought to be considered together with the outbreak of brilliancy of both orders on that day, as well as the apparently sudden existence of new ones.

Observation by Mr. Hirst of a dark shade on the moon.

The 'Observatory,' March 1, 1879, p. 375, contains an account, by Mr. H. C. Russell, of some Astronomical Experiments made on the Blue Mountains, near Sydney, N. S. W. Among these it is noticed that on 21st October, 1878, at 9 A.M., when looking at the moon, Mr. Hirst found that a large part of it was covered with a dark shade, quite as dark as the shadow of the earth during an eclipse of the moon. Its outline was generally circular, and fainter near the edges. Conspicuous bright lunar objects could be seen through it; but it quite obliterated the view of about half the moon's terminator, while those parts of the terminator not in the shadow were distinctly seen.

No change in the position of the shade could be detected after three hours' watching. The observation is made, "One could hardly resist the conviction that it was a shadow; yet it could not be the shadow of any known body. If produced by a comet, it must be one of more than ordinary density, although dark bodies have been seen crossing the sun which were doubtless comets." The diameter of the shadow from the part of it seen on the moon was estimated at about three quarters that of the moon[11].

---

[11] Some doubt has been cast on this observation, on the ground that nothing unusual was seen, and that the appearances were only those ordinarily presented by the moon at its then phase. I simply give the account as it appears in the scientific journal in which it was published.

CHAPTER VIII.

AURORA AND THE SOLAR CORONA.

Aurora and the solar corona. Mr. Norman Lockyer's 'Solar Physics.'

Mr. Norman Lockyer, in his 'Solar Physics,' a work of 666 pages, gives but little space to the Aurora. The index comprises:—"Aurora Borealis, connexion with sun-spots, pp. 82-102." "Affirmed coincidence of spectrum with that of the corona, pp. 244, 256."

Extracts from as to Aurora's connexion with sun-spots and with solar corona.

Page 82. After referring to Gen. Sabine as having shown that there are occasional disturbances in the magnetic state of the earth, and that these disturbances have a periodical variation, coinciding in period and epoch with the variation in frequency and magnitude of the solar spots as observed by Schwabe, the author proceeds to state, "and the same philosopher has given us reason to conclude that there is a similar coincidence between the outburst of solar spots and of the Aurora Borealis."

Page 102. "We have also shown that sun-spots or solar disturbances appear to be accompanied by disturbances of the earth's magnetism, and these again by auroral displays."

Evidence of American observers on nature of the corona considered.

Page 243. "What, then, is the evidence furnished by the American observers on the nature of the corona (solar)? It is bizarre and puzzling to the last degree. The most definite statement on the subject is that it is nothing more nor less than a *permanent Solar Aurora!* the announcement being founded on the fact that three bright lines remained visible after the image of a prominence had been moved away from the slit, and that one (if not all) of these lines is coincident with a line (or lines) noticed in the spectrum of the Aurora Borealis by Professor Winlock." Mr. Lockyer then adds, that amongst the lines he had observed up to that time, some forty in number, this line was among those which he had most frequently recorded, and was, in fact, the first iron line which made its appearance in the part of the spectrum he generally studied, when the iron vapour is thrown into the chromosphere.

Mr. Lockyer's conclusion adverse to the question being settled.

Hence he thought he should always see it if the Aurora were a permanent solar corona, and gave out this as its brightest line, and on this ground alone should hesitate to regard the question as settled.

Prof. Young's communication to 'Nature.'

Page 256 is an extract from a communication by Prof. Young to 'Nature,' March 24, 1870, in which the Professor refers to the bright line 1474 as being always visible with proper management. He also thinks it probable that this line coincides with the Aurora line reported by Prof. Winlock at 1550 of Dr. Huggins's scale, though he is by no means sure of it. He had only himself seen it thrice, and then not long enough to complete a measurement. He was only sure that its position lay between 1460 and 1490 of Kirchhoff.

He does not abandon his hypothesis, it having other elements of probability.

For this reason he did not abandon the hypothesis, which appeared to have other elements of probability, in the general appearance of the corona, the necessity of immense electrical disturbances in the solar atmosphere as the result of the powerful vertical currents known to exist there, as well as the curious responsiveness of our terrestrial magnets to solar storms; yet he did not feel in a position to urge it strongly, but rather awaited developments. Father Secchi was disposed to think the line hydrogen, while Mr. Lockyer still believed it to be iron.

Dr. Schellen reviews the subject in eclipse of 1869.

Dr. Schellen, in his 'Spectrum Analysis,' treats the matter more in detail. Referring to the eclipse of 1869 as confirming the previous observations that the coronal spectrum was free from dark lines, he points out that Pickering, Harkness, Young, and others were agreed that with the extinction of the last rays of the sun all the Fraunhofer lines disappeared at once from the spectrum. He further says:—

Young observed three bright lines in the spectrum of the corona. Coincidence of these lines with three bright lines observed by Winlock in the Aurora. Corona self-luminous, and probably of a gaseous nature. Corona supposed to be a permanent polar light existing in the sun. Polar light in the sun attributed to electricity. Dr. Schellen thinks nature of the corona still a problem.

"The small instruments employed by Pickering and Harkness, with a large field of view, exhibited a spectrum obtained at once from the corona, the prominences, and the sky in the neighbourhood of the sun. These instruments showed during totality a faint continuous spectrum free from dark lines, but crossed by two or three bright lines. Young, with a spectroscope of five prisms, observed the three bright lines in the spectrum of the corona, and deduced the following positions according to Kirchhoff's scale:—1250 ± 20, 1350 ± 20, and 1474. It had been already explained why the last and brightest of these lines was thought to belong to the corona and not to that of the prominences, and it seemed probable that the other two lines belonged also to the light of the corona, from the fact that they were both wanting in the spectrum of the prominences when observed without an eclipse. But what invested these three lines with a peculiar interest was the circumstance that they appeared to coincide exactly with the first three of the five bright lines observed by Professor Winlock in the spectrum of the Aurora Borealis. These lines of the Aurora were determined by Winlock according to Huggins's scale; and if these be reduced to Kirchhoff's scale, the positions of the lines would be 1247, 1351, and 1473, while the lines observed by Young were 1250, 1350, and 1474." Dr. Schellen then points out that if it be borne in mind that Young found the positions of the two fainter lines more by estimation than by measurement, the coincidence between the bright lines of the corona and those of the Aurora would be found very remarkable. The brightest of the lines, 1474, was the reversal of a strongly marked Fraunhofer line, ascribed by Kirchhoff and Ångström to the vapour of iron. Dr. Schellen then details Professor Pickering's observations with the polariscope, showing that the corona must be self-luminous, and that from the bright lines seen in its spectrum it is probably of a gaseous nature, and forms a widely diffused atmosphere round the sun; and then adds, "It has been supposed, from the coincidence of the three bright lines of the corona with those of the Aurora Borealis, that the corona is a permanent polar light existing in the sun analogous to that of our earth." Dr. Schellen here adds:—"Lockyer, however, justly urges against this theory the fact that although the brightest of these three lines, which is due to the vapour of iron, is very often present among the great number of bright lines occasionally seen in the spectrum of the prominences, it is by no means constantly visible, which ought to be the case if the corona were a permanent polar light in the sun." (Professor Young's answer to this, on the ground of line 1474 being always visible, has been already given.) "A yet bolder theory is the ascription of such a polar light in the sun to the influence of electricity, which has been proved, it is well known, by the relation of the magnetic needle, and the disturbance of the electric current in the telegraph wires, to play an important part in the phenomenon of the Aurora Borealis;" and Dr. Schellen then concludes with an opinion that the nature of the corona was still a problem[12].

Various places of wave-length assigned to these lines.

On reference to the 'American Journal of Science,' vol. xlviii. pp. 123 and 404, it seems that the auroral observations before referred to were made on 15th April, 1869, by C. S. Pierce, with "an ordinary chemical spectroscope, with the collimator pointed directly to the heavens," and were reported by Winlock. The lines were 1280, 1400, and 1550 of Huggins's scale, and were reduced to Kirchhoff's scale by Young. These lines have had all sorts of places of wave-length assigned to them by different writers. Proctor gives 5570, 5400, 5200; Pickering and Alvan Clarke, 5320 (assumed to be 5316, coronal line); Barker, 5170, 5200, 5020; Backhouse, 5320, 4640, and 4310. In my 'Aurora Spectrum,' Plate XII., I have assigned two, with a?, to 5320 (Alvan Clarke) and 5020 (Barker). The third might perhaps be placed at 4640 (Backhouse and Winlock).

---

[12] The question of a connexion between the waxing and waning of the solar corona and the prevalence of sun-spots is now being mooted, and may have an important bearing on the subject of the constitution of the corona. It would seem that when the corona has been examined about the time of minimum of sun-spots, it has proved fainter though more extended, while the bright lines of the spectrum have been absent, indicating a change or variance in the gaseous part of it at those periods.

Doubts raised as to closeness of the observations for the purpose of comparison.

The coincidences relied on in the foregoing observations depend, of course, upon (1) the accuracy of the observations themselves, and (2) the subsequent reduction of the lines for comparison. Assuming the correctness of the latter, what have we as to the former? Two of Professor Young's positions of coronal lines, as stated, seem to have far too much of the ± element to make them sufficiently accurate. Pierce's auroral observation does not state how the lines were positioned. As they *all* end with a cypher, the suspicion naturally arises that the measurements did not extend beyond the first three places of the figures, and, if so, could not be used for accurate comparison. The auroral lines, too, are generally rather wide and nebulous, and not easy of comparison with sharper ones.

CHAPTER IX.
SUPPOSED CAUSES OF THE AURORA.

Supposed causes of the Aurora. Sulphurous vapours. Magnetic effluvia.

At first the Aurora was described to be sulphurous vapours issuing from the earth; and Musschenbroek pointed out that certain chemical mixtures sent forth a phosphorescent vapour, in some respects resembling the Aurora. Dr. Halley originally proposed a similar theory, but ultimately concluded that the Aurora might be occasioned by the circulation of the magnetic effluvia of the earth from one pole to another.

Zodiacal light.

M. de Mairan, in 1721, in a treatise, ascribed the Aurora to the impulse of the zodiacal light upon the atmosphere of the earth.

Luminous particles of our atmosphere.

Euler combated this theory, and ascribed the Aurora to the luminous particles of our atmosphere driven beyond its limits by the light of the sun, and sometimes ascending to the height of several thousand miles.

Electric fluid in vacuo resembles Aurora.

Mr. Hawksbee very early showed that the electric fluid assumes, *in vacuo* or in highly rarefied atmosphere, an appearance resembling the Aurora. Mr. Canton contrived an imitation of the Aurora by means of electricity transmitted through the Torricellian vacuum in a long glass tube, and showed that such a tube would continue to display strong flashes of light for 24 hours and longer without fresh excitation.

Experiment with electrical machine and exhausted receiver.

In the 'Edinburgh Encyclopædia,' date 1830, is mentioned an experiment in which an electrical machine and air-pump are so disposed that strong sparks pass from the machine to the receiver of the air-pump.

Dr. Franklin's theory.

As the exhaustion proceeds the electricity forces itself through the receiver in a visible stream, at first of a deep purple colour; "but as the exhaustion advances it changes to blue, and at length to an intense white, *with which the whole receiver becomes completely filled."* [It will be noticed that this experiment bears a close resemblance to Prof. Ångström's exhausted flask referred to later in treating of the spectrum of the Aurora.]

Dr. Franklin gave a different form to the electric theory of the Aurora, supposing that the electricity which is concerned in the phenomenon passes into the Polar regions from the immense quantities of vapour raised into the atmosphere between the tropics (Exper. and Observ. 1769, p. 43).

Mr. Kirwan's theory.

Mr. Kirwan (Irish Trans. 1788) supposed that the light of the Aurora Borealis and Australis was occasioned by the combustion of inflammable air kindled by electricity.

Mons. Monge's.

Mons. Monge proposed the theory that the Auroræ were merely clouds illuminated by the sun's light falling upon them after numerous reflections from other clouds placed at different distances in the heavens (Leçons de Physique par Prejoulz, 1805, p. 237).

Mons. Libes'.

Mons. Libes propounded a theory that the electric fluid, passing through a mixture of azote and oxygen, produced nitric acid, nitrous acid or nitrous gas, and that these substances, acted upon by the solar rays, would exhibit those red and volatile vapours which form the Aurora Borealis (Traité de Physique, ou Dictionnaire de Physique, par Libes; Rozier's Journal, June 1790, February 1791, and vol. xxxviii. p. 191).

Mr. Dalton's.

Mr. Dalton considered the Aurora a magnetic phenomenon whose beams were governed by the magnetism of the earth. He observed that the luminous arches were always perpendicular to the magnetic meridian (Dalton's Meteorological Observations and Essays, 1793, pp. 54, 153).

Abbé Bertholon's.

The Abbé Bertholon ascribed the Aurora Borealis to a phosphorico-electric light (Encyc. Méthod. art. Auroræ).

Dr. Thompson concluded the arches to be an optical deception.

Dr. Thompson (Annals of Philosophy, vol. iv. p. 429), from the observations of Mr. Cavendish and Mr. Dalton, concluded there was no doubt that the arched appearance of the Aurora was merely an optical deception, and that in reality it consisted of a great number of straight cylinders parallel to each other and to the dipping-needle at the place where they were seen.

Artificial Auroræ produced in exhausted tubes.

With many of us (at least it was so in my own case) our first viewed Auroræ have been artificial ones, devised by electricians and having their locus at the Royal Polytechnic in Regent Street or in some scientific lecture-room. The effects in these cases are produced in tubes nearly exhausted by means of an air-pump, and then illuminated by some form of electric or galvanic current.

Tubes described.

In one instance the tube is usually of the form shown on Plate X. fig. 9, supported on a base with a brass ball electrode at the lower end, and a pointed wire at the upper. In another case the tube is of the form shown on same Plate, fig. 8. After exhaustion it is permanently closed, the current passing through it by means of the platinum-wire electrodes introduced into each end of the tube. The first form of tube is usually excited by a frictional plate machine; the second by a galvanic current from a Grove or bichromate battery, which, by the aid of a Ruhmkorff coil, has had its character changed from quantity to intensity. In each instance, upon connexion with the source supply of the electric current, a very similar effect is produced.

Effects described.

Brilliant streams of rose-coloured light pass between the electrodes, sometimes as a single luminous misty band, sometimes in divided vibrating sprays or streams, and sometimes in a flaky column of striæ.

All this, before the spectroscope took its part in the investigation, we were content to accept as a very fair and probable explanation of the Aurora accompanied by a mimic representation of the phenomenon.

These appearances may, of course, be produced at will in tubes having electrodes; but it is, moreover, possible to produce them, though with less effect, in certain other forms of tube having no such direct communication with the external electric machine.

One electrode only may be connected with the coil or electrical machine. The appearance is then a faint representation of what happens when the current entirely passes (but see experiments with a single wire detailed in Part III.).

Tube without electrodes.

In the case of an exhausted tube having no electrodes, the wires from the coil may be made into a little helix and placed at each end of the tube, and the induced currents within will show themselves in flashes and streams of light, varying in colour and tint according to the gaseous or other contents of the tube.

Tube excited by friction.

In some cases the ordinary forms of galvanic or electrical machine for supplying the current of electricity may be dispensed with. A long straight tube exhausted and closed at each end, and without electrodes, Plate X. fig. 6, being slightly warmed and then excited by friction with the dry hand or a piece of flannel, silk handkerchief, or the like, is soon filled with the most brilliant flashes of light playing in the interior, and when once thoroughly charged needs but little further excitation to keep up the effect.

Geissler's mercury tube.

Geissler has introduced a form of tube in which electricity in its form of flashes and glow of light is produced by the friction of mercury. The outer tube is strong, and contains within it a smaller tube of uranium glass with balls blown upon it (Plate X. fig. 7). The tubes are exhausted and a small quantity of mercury is introduced which has access to both surfaces of the inner tube, as well as to the inner surface of the outer tube. Upon the tube being reversed end for end or shaken, the mercury runs up or down the tube and causes a very considerable display of whitish light.

The before-described tubes are also referred to, and their spectra described, in the section "On the comparison of some tube and other Spectra with the Aurora" (Part II.).

The aura or brush from the electrical machine has been considered as resembling the Aurora, while the hissing and crackling accompanying it has been supposed to corroborate the reports of similar noises having been heard during an auroral display.

Prof. Lemström's instrument to demonstrate the nature of Auroræ.

Prof. Lemström, of the University of Helsingfors, has devised an instrument for the purpose of demonstrating that Auroræ are produced by electrical currents passing through the atmosphere. An illustration of this instrument (for which I am indebted to the Editor of 'Nature') is introduced (fig. 1).

The instrument was exhibited at the recent Scientific Loan Collection at South Kensington, and a full description of it, together with an essay by Prof. Lemström, "On the Theory of the Polar Light," will be found in the third edition of the Official Catalogue, p. 386. no. 1751. The apparatus is intended to show that an electric current passing from an insulated body does not produce light in air of normal pressure; but as it rises to the rarefied air in the Geissler tubes a phenomenon very like the real Polar Light is produced.

Fig. 1.

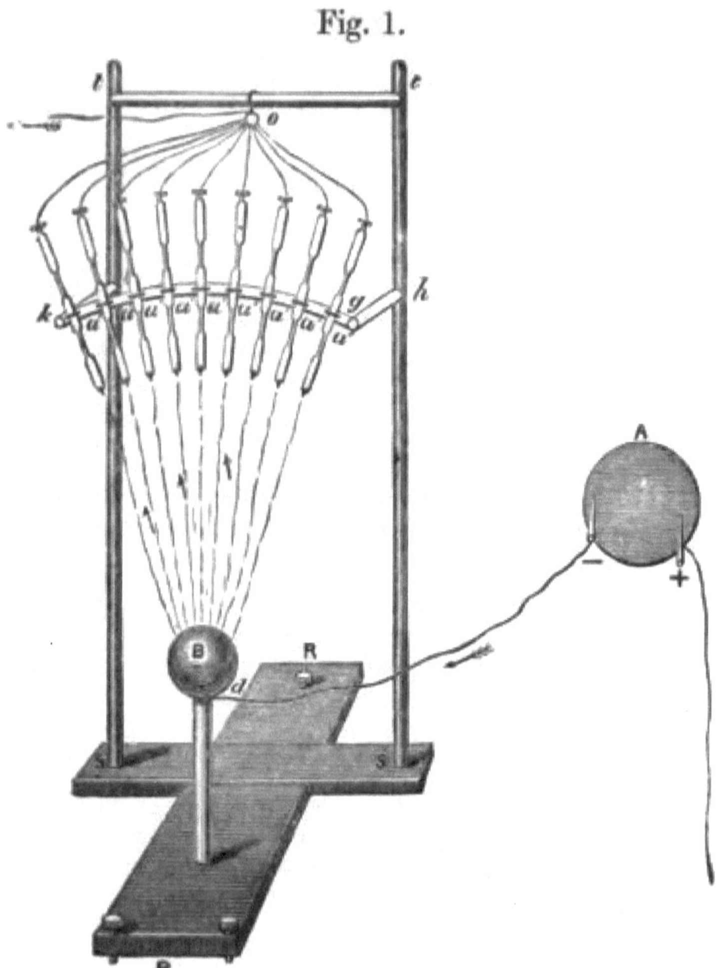

Fig. 1.

A is an electrical machine, the negative pole being connected with a copper sphere and the positive with the earth.

$s\,s'$ is of ebonite as well as R R d, so that B is quite insulated as the earth is in space. B is surrounded by the atmosphere. $a'\,a'\,a'\,a'\,a'\,a'$ are a series of Geissler tubes with copper ends above and below. All the upper ends are connected with a wire which goes to the earth; consequently a current runs in the direction of the arrows through the air, and the Geissler tubes become luminous when the electrical machine is set into operation.

The Geissler tubes represent the upper part of the atmosphere which becomes luminous when the Aurora Borealis is observed in the northern hemisphere. The phenomena produced by the Lemström apparatus are considered consistent with the theory advocated by Swedish observers that electrical currents emanating from the earth and penetrating into the upper regions produce Auroræ in both hemispheres. The experiment differs from the apparatus of M. de la Rive, who placed his current *in vacuo*, and did not show the property of ordinary atmospheric air, in allowing to pass unobserved, at the pressure of 760 millims., a stream of electricity which illuminates a rarefied atmosphere.

M. de la Rive's apparatus described.

De la Rive's apparatus was also exhibited at the same time, and will be found described at p. 385 of the Catalogue, No. 1749. A large sphere of wood represented the earth, and iron cylinders the two extremities of the terrestrial magnetic axis. These penetrated into two globes filled with rarefied air, simulating the higher regions of the Polar atmosphere. The electric discharge turned around a point situate in the prolongation of the axis, in a different direction at either pole, when the two cylinders were charged by means of a horseshoe electro-magnet, in accordance with observations on the rotation of the rays of the Aurora.

De la Rive's magnet in an electric egg.

De la Rive placed an electro-magnet in an electric egg. As soon as the magnet was set in action the discharge which had before filled the egg was concentrated into a defined band of light, which rotated steadily round the magnet.

Gassiot's experiment with 400 Grove cells and exhausted receiver between poles of magnet.

Gassiot describes an experiment with his great Grove's battery of 400 cells, in which an exhausted receiver was placed between the poles of the large electro-magnet of the Royal Institution.

"On now exciting the magnet with a battery of 10 cells, effulgent strata were drawn out from the positive pole, and passed along the under or upper surface of the receiver according to the direction of the current.

"On making the circuit of the magnet and breaking it immediately, the luminous strata rushed from the positive pole and then retreated, cloud following cloud with a deliberate motion, and appearing as if swallowed up by the positive electrode." Mr. Marsh considered this bore a very considerable resemblance to the conduct of the auroral arches, which almost invariably drift slowly southward.

Mr. Marsh considered the Aurora an electric discharge between the magnetic poles of the earth.

He considered it probable that the Aurora was essentially an electric discharge between the magnetic poles of the earth, leaving the immediate vicinity of the north magnetic pole in the form of clouds of electrified matter, which floated southward, bright streams of electricity suddenly shooting forth in magnetic curves corresponding to the points from which they originated, and then bending southward and downward until they reached corresponding points in the southern magnetic hemisphere, and forming pathways by which the electric currents passed to their destination; and, further, that the magnetism of the earth caused these currents and electrified matter composing the arch to revolve round the magnetic pole of the earth, giving them their observed motion from east to west or from west to east.

Varley's observation on a glow-discharge in vacuo. Spark surrounded by an aura which could be separated.

Varley showed that when a glow-discharge in a vacuum tube is brought within the field of a powerful magnet, the magnetic curves are illuminated beyond the electrodes between which the discharge is taking place, as well as in the path of the current, and also thought that this illumination was caused by moving particles of matter, as it deflected a balanced plate of talc on which it was caused to infringe. It has also been shown that in electrical discharges in air at ordinary pressure, while the spark itself was unaffected by the magnet, it was surrounded by a luminous cloud or aura which was driven into the magnetic curve, and which might also be separated from the spark by blowing upon it.

Most of the foregoing interesting results and experiments will be found repeated and verified in Part III.

Prof. Lemström's Theory.

Prof. Lemström's theory.

Prof. Lemström thinks that terrestrial magnetism plays only a comparatively secondary part in the phenomena of the Polar Light, this part consisting essentially in a direct action upon the rays.

That the experiments of M. de la Rive do not all furnish the proof that the rays of the light are really united under this influence.

Character of the Polar Light.

That the Polar Light considered as an electrical discharge gives the following results:—

(1) An electric current arising from the discharge itself, which takes place slowly.

(2) Rays of light consisting of an infinite number of sparks, each spark giving rise to two induction currents going in opposite directions.

(3) A galvanic current going in an opposite direction to that of the discharge, and having its origin in the electromotive force discovered by M. Edlund in the electric spark. That these currents require a closed circuit; but this is not necessary in the case of the Aurora, as the earth and rarefied air of the upper regions are immense reservoirs of electricity producing the same effect as if the circuit were closed. That permanent moisture in the air, a good conductor of electricity, is the cause of a slow and continuous discharge assuming the form of an Aurora, instead of suddenly producing lightning as in equatorial regions and mean latitudes.

Polar Light due to electric discharges only.

He sums up, that the electric discharges which take place in the Polar regions between the positive electricity of the atmosphere and the negative electricity of the earth are the essential and unique cause of the formation of the Polar Light, light the existence of which is independent of terrestrial magnetism, which contributes only to give to the Polar Light a certain direction, and in some cases to give it motion.

This Prof. Lemström maintains contrary to those who believe they see in terrestrial magnetism, or rather in the induction currents, what is capable of developing the origin of the Polar Light.

Theories of MM. Becquerel and De la Rive.

Theories of MM. Becquerel and De la Rive.

M. Becquerel's theory is that solar spots are cavities by which hydrogen and other substances escape from the sun's protosphere. That the hydrogen takes with it positive electricity which spreads into planetary space, even to the earth's atmosphere and the earth itself, always diminishing in intensity because of the bad conducting-power of the successive layers of air and of the earth's crust. That would then only be negative, as being less positive than the air. The diffusion of electricity through planetary space would be limited by the diffusion of matter, since it cannot spread in a vacuum. That gaseous matter extends further than the limits usually assigned to the earth's atmosphere, is proved by the observation of Auroræ at heights of 100 and 200 kilometres, where some gaseous matter must exist. M. de la Rive agrees with M. Becquerel as to the electric origin of the Aurora, but considers the earth is charged with negative electricity and is the source of the positive atmospheric electricity, the atmosphere becoming charged by the aqueous vapour rising in tropical seas. The action of the sun he considers is an indirect one, varying with the state of the sun's surface, as shown by coincidences in the periods of Aurora and sun-spots.

M. Planté's Electric Experiments.

M. Planté's experiments. Effects produced resembling Auroræ.

M. Planté has performed some experiments with a very considerable series of secondary batteries. By inserting the positive electrode after the negative in a vessel of salt water, luminous and other effects were observed which were considered to have a strong resemblance to those of Auroræ.

M. Planté advocates the theory that the imperfect vacuum of the upper regions, acting like a large conductor, plays the part of the negative electrode in his experiments, while the positive electricity flows towards the planetary spaces, and not towards the ground, through the mists and ice-clouds which float above the Poles.

M. Planté's experiments producing a corona, an arc, or a sinuous line.

In an article in 'Nature,' March 14, 1878, a further account is given of M. Planté's experiments, under the head of "Polar Auroræ;" and it is stated that, in these experiments, the electric current, in presence of aqueous vapour, yielded a series of results altogether analogous to the various phases of Polar Auroræ. If the positive electrode of the secondary battery was brought into contact with the sides of a vessel of salt water, there was observed, according to the distance of the film (electrode?) from the liquid, either a corona formed of luminous particles arranged in a circle round the electrode (fig. 2, p. 90), an arc bordered with a fringe of brilliant rays (fig. 3), or a sinuous line which rapidly folded and refolded on itself (fig. 4). This undulatory movement, in particular, formed a complete analogy with what had been compared in Auroræ to the undulations of a serpent, or to those of drapery agitated by the wind. The rustling noise accompanying the experiment was analogous to that sometimes said to accompany Auroræ, and was caused by the luminous electric discharge penetrating the moisture. As in Auroræ, magnetic perturbations were produced by bringing a needle near the circuit, the deviation increasing with the development of the arch.

The Auroræ were produced by positive electricity, the negative electrode producing nothing similar.

Illustrations of these miniature Auroræ are given in 'Nature,' and reproduced on p. 90. No mention of any spectroscopic observations is made.

Mr. Holden's views.

In a communication to the Metropolitan Scientific Association ('Observatory,' March 1, 1879, p. 389), Mr. A. P. Holden, after supporting the theory of a connexion between the waxing and waning of the solar corona and sun-spots, adopts Mr. F. Pratt's hypothesis "that the Aurora is simply light filmy cirrus cloud, first deposited at the base of a vast upper body of highly rarefied vapour, and illuminated by the free electricity escaping in the condensation through the very rarefied medium above, towards the north or south. The Aurora would, according to this theory, have its origin in a vast electrical storm, resulting from a violent condensation of vapour which causes a flow of electricity from the pole to restore equilibrium." The Aurora would thus, in Mr. Holden's opinion, "depend on storm phenomena of an intense character; and the frequency of Auroræ at the sun-spot maxima would indicate the connexion of the latter with the weather."

Fig. 2.

The corona.

Fig. 3.

The arc and rays.

Fig. 4.

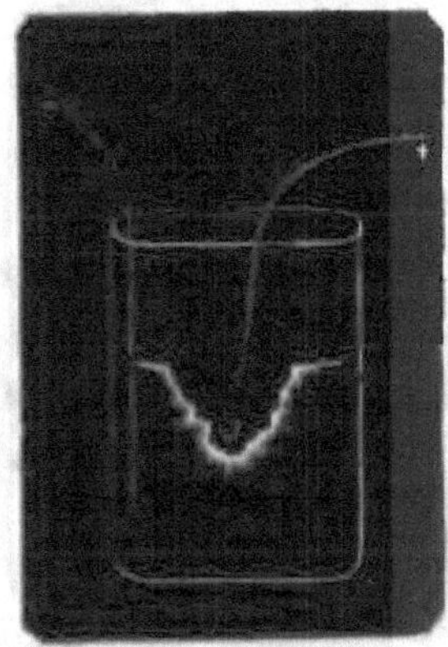

The sinuous line.

Fig. 2. The corona.
Fig. 3. The arc and rays.
Fig. 4. The sinuous line.

PART II.
THE SPECTRUM OF THE AURORA.

CHAPTER X.

SPECTROSCOPE ADAPTED FOR THE AURORA.

Must be of moderate dispersion, with ready mode of measuring line-positions.

Any form of spectroscope of moderate dispersion will suffice for observations of the spectrum of the Aurora; but, for sake of convenience, a hand or direct-vision spectroscope is to be preferred, and it is desirable also to have some quick and ready mode of measuring the position of the lines while the Aurora lasts.

Mr. Browning's instrument described.

Mr. John Browning arranged for me a form of instrument which I have found very convenient for observations by hand of the Aurora-lines, and also, when fixed on a stand, for tube and chemical investigations. A representation of this instrument is given on Plate X. fig. 1. A brass tube carries a large compound (5) direct-vision prism (shown dark in the drawing). An arrangement is made so that a second prism can at will be slipped into the tube (shown in outline in the drawing). With one prism and a fine slit the D lines are widely separated, and the field of view extends at one glance from near C to near G. When the second prism is inserted and used in combination, the nickel line can be seen between the two D lines, and the instrument may be used for solar work. A photograph of the sun's spectrum, taken with one prism only, shows a great number of the dark solar lines and many of the bright ones, ascribed by Prof. Draper to oxygen and nitrogen.

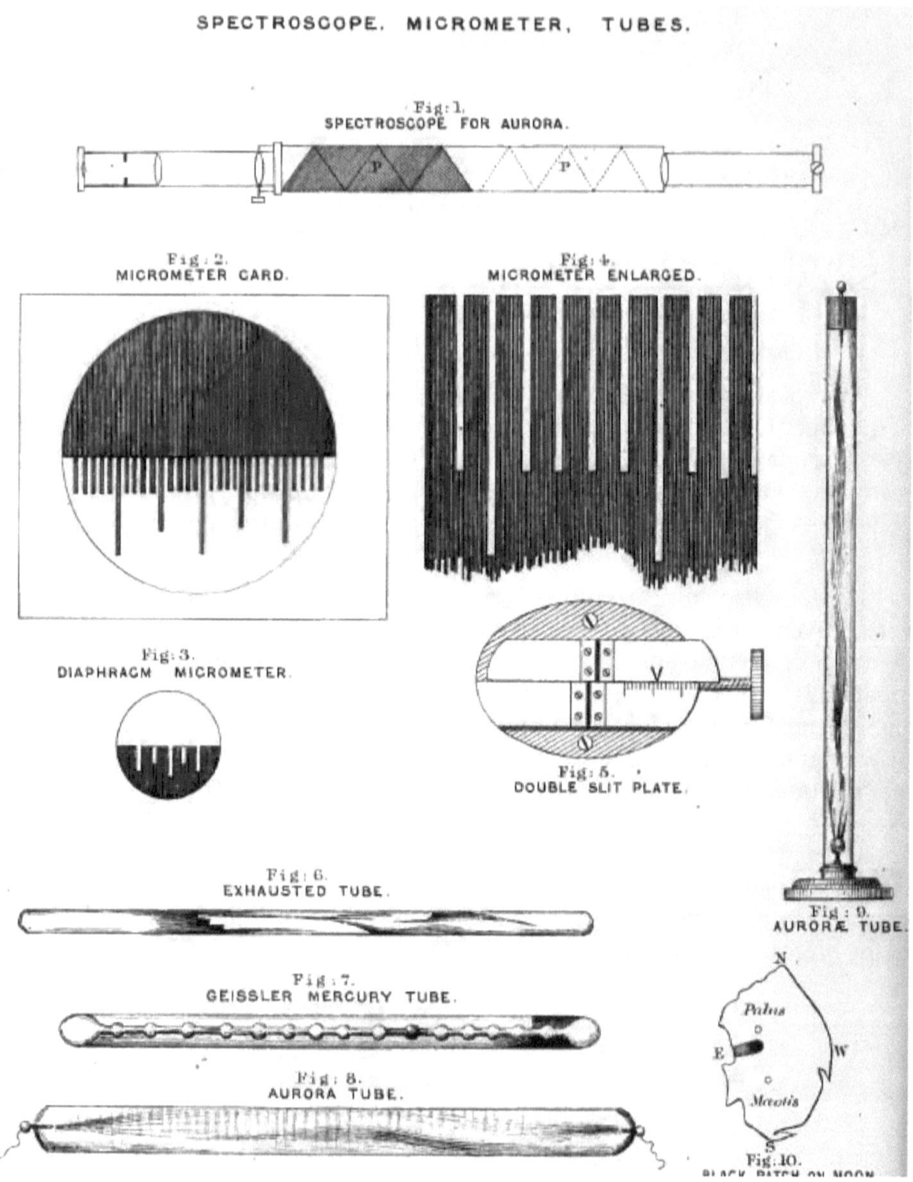

Plate X.

SPECTROSCOPE, MICROMETER, TUBES

Diaphragm micrometer described. Mode of use of the micrometer.

The collimator and observing telescope are respectively 6 inches in length, and carry achromatic lenses of one inch aperture. The telescope traverses the field so that the extremities of the spectrum may be observed. The dispersion of the instrument was ascertained by a set of observations of the principal solar and some metallic lines, made with an excellent filar micrometer. For the Auroral observations, Dr. Vogel has described an instrument (see Appendix E) in which the usual spider's-web wires are replaced by a needle-point, as being easily seen upon a faint spectrum. Illuminated wires may also be used; but I was led ultimately to employ, in preference, a diaphragm micrometer which the spectrum itself illuminates, as being adapted for speedy, yet fairly accurate, observations. It was made in this manner:—A card was first of all prepared (Plate X. fig. 2), and within a circle described on this, a scale was drawn of moderately wide white spaces, with black divisions between, short and long, so as to read off easily by eye. The upper half of the circle was then entirely filled in with black; and from the card as thus prepared a reduced negative photograph was made. In this the spaces and lower half of the circle were opaque, and the upper half of the circle and the lines between the spaces were transparent (Plate X. fig. 4). This photograph was about the size of a shilling (fig. 3, same Plate). It was mounted carefully in Canada balsam, with a thin glass cover, and then placed in the focus of the eyepiece. In use, the spectrum is brought upon the scale so that the upper half shows above the scale without any interruption at all; while the lower half illuminates the scale and renders the divisions visible, showing the spectrum-lines falling either upon them or the spaces between. The photographed scale was next enlarged to a considerable size and printed upon faintly ruled paper; and the enlargement was so arranged as to comprise five of the faint ruled lines between each division of the scale. Each of these faint lines in turn represented a certain portion of the spectrum as read off with the filar micrometer; so that the scales as constructed with the filar micrometer and with the photographed micrometer corresponded for all parts of the spectrum included in the field of the eyepiece.

Advantage of the method.

One of the photographed enlargements being laid on the table under the spectroscope, the observed lines were marked off with ease and accuracy upon it; and as the photograph was an exact copy of the scale, any want of exactitude in the divisions was of no moment.

One great advantage of this method was, that all the lines seen could be recorded at one time and with all in view, and without the risk of slight shift in the instrument, which frequently happens when lines are read off seriatim.

I found this plan most effective for the rapid and correct recording of a faint and evanescent spectrum, and it gave close results when compared with traversing-micrometer measured spectra. The records, too, admitted of subsequent examination at leisure.

Double-slit plate arrangement.

Mr. Browning subsequently constructed for me a double-slit plate (lately in the Scientific Loan Collection at South Kensington) for the same instrument (Plate X. fig. 5). The lower half of the plate is fixed. The upper half traverses the lower by the aid of a micrometer-screw. The slit is widened or closed at pleasure by loosening the small screws by which the jaw-plates are attached. A scale is engraved on the fixed lower half of the plate for an approximate measurement, while the division of the micrometer-screw-head completes it.

In use, one half of the spectrum slides along the other, and a bright line in the upper spectrum is selected as an index. The distances between the lines of the lower half of the spectrum are read off by means of the bright line above. This form of micrometer was suggested by Mr. Procter (in 'Nature') many years ago as a substitute for a more complicated apparatus by Zöllner. Other instruments on a similar principle have been lately introduced, but for Aurora purposes I prefer a fixed scale.

Photographed spectrum suggested.

In 'Photographed Spectra' I have pointed out that we shall probably obtain no spectrum of the Aurora to be absolutely depended upon for comparison with other spectra until we succeed in a photographed one. From experiments made with a special prism of the Rutherfurd form, constructed for me by Mr. Browning (with which many gas-spectra have been already photographed), I see no reason, should an unusually bright Aurora favour us with a visit, why its spectrum may not be recorded in a permanent form, and with lines sufficiently well marked to be compared with other spectra. Rapid dry plates would be especially useful for such a purpose, and some Auroræ, if wanting in brilliancy, would doubtless compensate by their period of endurance.

Mr. Hilger's half-prism spectroscope.

Mr. Adam Hilger has also made for me one of his "half-prism" spectroscopes, in which considerable dispersion is obtained with but very little loss of light. This instrument has a simple and rapid micrometer arrangement, with a bright line as an index. I have (for want of Auroræ) had no opportunity of trying it, but I doubt not it is well adapted for such a purpose.

Spectrum of the Aurora described.

Lines or bands and continuous spectrum.

The spectrum of the Aurora consists of a set of lines or bands upon a dark ground at each extremity of the spectrum, but with more or less of faint continuous spectrum towards the centre. The extreme range of the spectrum, as observed up to the present time, is from "*a*" (between C and D) in the red to "*h*" (hydrogen) in the violet.

Lines nine in number.

The lines have been classified and arranged by Lemström and others as nine in number, but I believe not more than seven have ever been seen simultaneously.

The author of the article "Aurora Polaris," in the 'Encyclopædia Britannica,' classes the lines as nine, and gives a table with the following results (to these I have added Herr Vogel's lines, for the purpose of identification and comparison):—

Table from Encyc. Brit.

No. of line.	Number of observations.	Mean W.L.	Probable error.	Vogel's lines.
1.	5	6303	± 8·1	6297
2.	10	5569	± 2·9	5569
3.	4	5342	±16	5390
4.	6	5214	± 5·4	5233
5.	4	5161	± 9·7	5189
6.	6	4984	±11	5004
7.	4	4823	± 9·3	
8.	8	4667	± 9·8	4663
9.	8	4299	± 9·3	

The probable errors are large, and it is a question whether any thing is gained by thus endeavouring to average the lines.

Ångström's line. Zöllner's line in the red. Other Lines of the spectrum.

The principal and brightest line, in the yellow-green, is generally called "Ångström's," and his (probably the first) measurement of its position at 5567 adopted. This was in the winter of 1867-68, and he saw in addition, by widening the slit, traces of three very feeble bands situated near to F. Zöllner is credited with the first observation of the line in the red. These two lines are generally described as with similar characteristics, and in about the same respective positions, by all observers, and have never been remarked to spread into bands. The other lines in the spectrum are difficult to position, owing to the many discordant observations of them. They seem also variable in intensity as well as in number (sometimes even in the same Aurora), and are not unfrequently observed to have their places supplied by bands.

Second German expedition observations. Austro-Hungarian.

The spectroscope was used in the second German expedition, but only the one brightest line seen—Dr. Börgen stating he had never seen a trace of the weak lines in the blue and red, which were observed so distinctly with the same spectroscope on 25th October, 1870, after the return of the expedition. Lieutenant Weyprecht used a small spectroscope during the Austro-Hungarian Expedition, and saw only the well-known yellow-green line.

Swedish expedition, 1868. Lemström's observations.

In the Swedish Expedition, 1868, Lemström mentions that in the Aurora spectrum there are nine lines (he does not say he saw them simultaneously), which he considers to agree with lines belonging to the air-gases. He also thinks the Aurora could be referred to three distinct types, depending on the character of the discharge.

Spectrum or Aurora seen at Tronsa.

At Tronsa an Aurora was seen October 21st, 1868, which commenced in the north and became very brilliant. The spectroscope showed:—

1. A yellow line at 74·9.

2. A very clear line in the blue at 65·90.

3. Two lines of hair's breadth, with very pronounced (horizontal?) striæ on the side of the yellow, one at 125 and the other about 105.

[I presume the striæ were really vertical, and that the explanation intended to convey that these lines shaded off towards the yellow. From a comparison of the figures they must have been in the red, and are the only instance recorded of two auroral lines in that region. They are subsequently spoken of as "shaded rays."—J. R. C.]

MM. Wijkander and Parent's observations.

M. Auguste Wijkander and Lieut. Parent, of the Swedish Expedition in 1872-73, under Professor Nordenskiöld, used a direct-vision spectroscope, with a micrometer-screw movement of the prisms, the reading being afterwards reduced to wave-lengths upon Ångström's line-values.

The following Table gives the results, with Dr. Vogel's lines added for the sake of comparison:—

Lines.	Observations, Wijkander.			Observations, Parent.			Mean of both.	Vogel.
	Number.	W.L.	Probable error.	Number.	W.L.	Probable error.		
..	..	..	..	..	..	..	..	6297
..	..	..	..	..	..	..	..	5569
(1)	5	5359	±3	..	..	..	5359	5390
(2)	6	5289	±5	3	5280	± 1	5286	..
(3)	6	5239	±4	2	5207	±11	5231	5233
..	..	..	..	..	..	..	..	5189
(4)	5	4996	±9	..	..	..	4996	5004
(5)	1	4871	..	1	4873	..	4872	..
(6)	8	4692	±2	10	4708	± 5	4701	4663
(7)	1	4366	..	..	..	..	4366	..
(8)	4	4280	±3	3	4286	±16	4284	..

The brightest line in all Auroræ, 5567, was intentionally not included in the Tables. The red line was not seen. Nos. 5 and 7 were only seen once, and not in the same Aurora.

Spectrum of Aurora of October 24th, 1870.

The Aurora of October 24th, 1870, came at a time when spectroscopes of a direct-vision form were being introduced, and a number of observations were communicated at the time to 'Nature.'

T. F.'s observations. W. B. Gibbs's observation. Elger's observation.

A correspondent, T. F., writing from Torquay, saw, with a direct-vision spectroscope, one strong red line near C, one strong pale yellow line near D, one paler near F, and a still paler one beyond, with a faint continuous spectrum from about D to beyond F. The C line was very conspicuous and the brightest of the whole. It was intermediate in position and colour to the red lines of the lithium and calcium spectra. Plainly there were two spectra superposed, for while the red portions of the Aurora showed the four lines with a faint continuous spectrum, the greenish portions showed only one line near D on a faint ground. W. B. Gibbs saw, in London, only two bright lines, one a greenish grey, situate about the middle of the spectrum, and the other a red line very much like C (hydrogen). Thomas G. Elger, at Bedford, on the 24th and 25th, saw:—(1) a broad and well-defined red band near C; (2) a bright white band near D (same as Ångström's W.L. 5567), on 25th visible in every part of the sky; (3) a faint and rather nebulous line, roughly estimated to be near F; (4) a very faint line about halfway between 2 and 3. The red band was absent from the spectrum of the white rays of the Aurora, but the other lines were seen.

J. R. Capron's observation.

With a small Browning direct-vision spectroscope on the 24th, I found no continuous spectrum, but two bright lines, one in the green (like that from the nebulæ, but more intense, and considerably flickering), the other in the red (like the lithium line, but rather duskier: Plate V. fig. 6). The latter was only well seen when the display was at its height; it could, however, be faintly traced wherever the rose tint of the Aurora extended. The line in the green was well seen in all parts of the sky, but was specially bright in the Auroral patches of white light.

Mr. Browning's observation. Alvan Clarke's, jun., observations.

Mr. Browning also saw the red line, but found comparison difficult. On the evening of the 24th October, Mr. Alvan Clarke, jun., at Boston, used a chemical spectroscope of the ordinary form, with one prism and a photographed scale illuminated with a lamp. Four Auroral lines were seen at points of his scale numbered 61, 68, 80, and 98. These were reduced to wave-lengths by Professor Pickering, with the following results:—

Line.	Reading on scale.	Wave-lengths.	Assumed line.	Comments.	Probable error.
1.	61	5690	5570	Common Aurora-line.	-20
2.	68	5320	5316	Corona line.	+ 1
3.	80	4850	4860	F, hydrogen.	- 3

| 4. | 98 | 4350 | 4340 | G, hydrogen. | + 6 |

[61 is evidently wrong, and was probably a mistake for 63.]

G. F. Barker's observations.

George F. Barker, observing at New Haven (U.S.A.), saw, on November 9th, a crimson and white Aurora, which he examined with a single glass-prism spectroscope, by Duboscq, of Paris. The line positions were obtained by an illuminated millimetre scale. In the white Aurora were four lines (the red one being absent); in the red Aurora five. The wave-lengths of the Aurora-lines were run out as follows: —

(1.)	Between	C	and	D,	6230	(Zöllner's 6270).
(2.)	"	D	and	E,	5620	(Ångström's 5570).
(3.)	"	E	and	*b*,	5170	(Winlock's 5200).
(4.)	"	*b*	and	F,	5020	
(5.)	"	F	and	G,	4820	(Alvan Clarke's, jun., 4850).

Spectrum of Aurora of Feb. 4, 1872. Prof. Piazzi Smyth's observations.

Mr. Procter's Aurora-lines will be found noticed in connexion with the spectrum of oxygen; and Lord Lindsay's lines, with a comparison scale drawing, are separately described further on in this Chapter. The Aurora of February 4th, 1872, had many observers; some of whom communicated at the time spectroscopic notes. Professor Piazzi Smyth minutely describes the display as seen in Edinburgh, and saw "Ångström's green Aurora-line perpetually over citron acetylene[13] at W.L. 5579, and the red Aurora-line between lithium *a* and sodium *a*, but nearer to the latter, say at W.L. 6370." Extremely faint greenish and bluish lines also appeared at W.L. 5300, 5100, and 4900 nearly.

Rev. T. W. Webb's observations.

The Rev. T. W. Webb, with a very fine slit, saw the green Auroral line even in the light reflected from white paper. With a wider slit he saw a crimson band in the brighter patches of that hue, and beyond an extent of greenish or bluish light, which he suspected to be composed of contiguous bands.

R. J. Friswell's observations.

R. J. Friswell, coming up the Channel at 9.40, with a Hoffman's direct-vision spectroscope (the observing telescope removed), saw the green line, a crimson line near C, and faint traces of structure in the blue and violet.

---

[13] There seems to be some confusion as to the W.L. here given; 5567 is usually accepted as Ångström's line, while Prof. Smyth refers to it as 5579. The position, too, when examined with a spectroscope of greater dispersion, is not exactly over the citron-line of acetylene, both the above referred to lines lying somewhat more towards the violet end of the spectrum (see Plate V. fig. 7).

The Rev. S. J. Perry's observations.

The Rev. S. J. Perry observed at Stonyhurst four lines, and, on examining one of the curved streamers, found the red line even more strongly marked than the green. A magnetic storm was observed to be at its height from 4 to 9 P.M. of the same day.

J. R. Capron's observations.

With a Browning 7-prism direct-vision spectroscope I saw the green line in all parts of the Aurora, attended with a peculiar flickering movement. I did not see the other lines.

His catalogue of lines up to Nov. 9, 1872.

In a letter to 'Nature,' dated November 9th, 1872, I catalogued the lines observed up to that date as follows:—

1. A line in the red between C and D. W.L., Ångström, 6279.

2. A line (the principal one of the Aurora) in the yellow-green, between D and E. W.L., Ångström, 5567.

3. A line in the green, near E (corona line?). W.L., Alvan Clarke, jun., and Backhouse, 5320.

4. A faint line in the green, at or near *b*. W.L., Barker, 5170.

5. A faint line or band in the green, between *b* and F. W.L., Barker, 5020 (chromospheric?).

6. A line in the green-blue, at or near F. W.L., Alvan Clarke, jun., 4850.

7. A line in the indigo, at or near G. W.L., Alvan Clarke, jun., 4350.

8. The continuous spectrum from about D to beyond F.

Dr. H.C. Vogel's observations of Auroral lines. Spectrum described.

Dr. H. C. Vogel, formerly of the Bothkamp Observatory, near Kiel, and since of the Astrophysical Observatory, Potsdam, made several observations of the Auroral lines, October 25th, 1870. Besides the bright line between D and E, he found several other fainter lines stretching towards the blue end of the spectrum on a dimly-lighted ground. February 11th, 1871, he observed the same set of lines, and an average of six readings gave 5572 as the W.L. of the Ångström line. February 12th gave 5576 as Dr. Vogel's reading, and 5569 as Dr. Lohse's. April 9th gave 5569, and April 14th 5569. The Aurora of April 9th, 1871, was exceedingly brilliant, so that micrometer measurements of the lines were taken. The spectrum consisted of one line in the red, five in the green, and a somewhat indistinct broad line or band in the blue. The lines are thus described:—

Table of lines.

Table of Dr. Vogel's lines. Aurora, April 9th, 1871.

W.L.	Probable error.	Remarks.	
6297	14	Very bright stripe.	On a faintly lighted ground.
5569	2	Brightest line of the spectrum, became noticeably fainter at appearance of the red line.	
5390	..	Extremely faint line; unreliable observation.	
5233	4	Moderately bright.	
5189	9	This line was very bright when the red line appeared at the same time; otherwise equal in brilliancy with the preceding one.	
5004	3	Very bright line.	
4694		Broad band of light, somewhat less brilliant in the middle; very faint in those parts of the Aurora in which the red line appeared.	
4663	3		
4629			

A translation of Dr. Vogel's interesting paper will be found printed *in extenso* in Appendix E, and his lithographed drawings of the spectrum in the green and red portions of the Aurora respectively on Plate VI. figs. 2 and 3. The observations of April 9th by Dr. Vogel are probably, up to the present time, the most exact of any one Aurora, and I have therefore in most cases used them for comparison.

Mr. Backhouse's catalogue of lines.

Mr. Backhouse, in a letter to 'Nature,' commenting upon my catalogue of lines, gave the following as the latest determinations from his own observations:—

No. 1.	Wave-length	6060
2.	"	5660
3.	"	5165
4.	"	5015
6.	"	4625
7.	"	4305

(6060 must be a mistake for 6260, and 5660 for 5560.—J. R. C.) Mr. Backhouse never saw a line at 5320 again. He found the continuous spectrum to reach from No. 2 to No. 7, being brightest from a little beyond No. 2 to No. 6. This part of the spectrum did not give him so much the idea of a true "continuous spectrum" as of a series of bright bands too close to be distinguished.

Subsequent full catalogue of Auroral lines.

I have subsequently, in another section of this Chapter, added a full catalogue of the Auroral lines, prepared by myself from the foregoing and other sources and observations; and I also append to it a Plate [Plate XII.], in which these lines are positioned and the wave-lengths and names of observers are given. The numbers of the lines on the Plate correspond with those in the catalogue. The solar spectrum and the spectrum of the blue base of a candle-flame are added for purposes of comparison. [The telluric bands in the solar spectrum are shown more distinctly than they actually appear, and do not profess to give details.]

Flickering of the Green Line.

Flickering of the green line. Herschel's observation. J. R. Capron's observation.

A. S. Herschel noticed this, April 9th (1871?). He says:—"A remarkable circumstance connected with the appearance of the single line observed on this occasion was the flickering and frequent changes with which it rose and fell in brightness; apparently even more rapidly than the swiftly travelling waves, or pulsations of light, that repeatedly passed over the streamers, near the northern horizon, towards which the spectroscope was directed." In the spectrum of the Aurora of 20th October, 1870, I saw and noted the green line as "considerably flickering;" and in the Aurora of 4th February, 1872, I again saw and noted "the peculiar flickering" I had remarked in 1870. I have not seen the peculiarity noted by other observers.

Mr. Backhouse's graphical Spectra of four Auroræ.

Mr. Backhouse's graphical spectra of Auroræ.

Mr. Backhouse has been good enough to supply me with some details of four several Auroræ seen by him at Sunderland, accompanied by drawings, showing in a graphical way the spectrum of each display as seen with a spectroscope with rather a wide slit and as drawn by eye. I have reduced the four drawings to the same scale, and in this way they are extremely interesting for comparison (Plate V. fig. 4). The line on the left in each spectrum is Ångström's bright Auroral line, and is supposed to be considerably prolonged. The height of the lines denotes intensity.

April 18, 1873.

April 18th, 1873, was a bright Aurora. No. 3 is a faint band, which Mr. Backhouse had not perceived before. No. 5 had not been visible lately, and Mr. Backhouse thought it must belong to Auroræ of a different type from those which had appeared latterly.

Feb. 4, 1874.

February 4th, 1874. In the spectrum of this Aurora Mr. Backhouse saw seven lines, all that he had ever seen. (The red line, not shown in the diagram, makes the seventh.)

The spectrum is represented as seen between 6.50 and 7.5 P.M. Mr. Backhouse had only once before seen No. 4, and it became quite invisible between 7.45 and 7.55, though the other lines were as bright as before and the red line had appeared.

Oct. 3, 1874.

October 3rd, 1874. This spectrum was examined, and diagram made between 10 and 10.25 P.M. Five lines only are indicated.

It is mainly distinguished from the two preceding spectra by the brightness of the continuous spectrum on which the lines 2, 3, and 4 lie, and by the weakness of No. 5.

Oct. 4, 1874.

October 4th, 1874. Taken between 11.10 and 11.20 P.M.; distinguished, like the last, by a considerable amount of continuous spectrum and by a faint line (No. 3), not seen in the last spectrum, while No. 3 in the last is missing in this spectrum.

Mr. Backhouse's remarks as to comparative frequency of some of the Auroral lines.

Mr. Backhouse, as to both these last spectra, remarks that the lines were very variable in intensity, and sometimes some were visible and sometimes others. They varied also in relative brightness in different parts of the sky at the same time. Mr. Backhouse, in a communication to 'Nature,' referring to a statement of Mr. Procter's, that the bands of the Auroral spectrum are seldom visible, except the bright line at 5570, says that he always found two bands, "doubtless Winlock's 4640 and 4310," to be invariably visible when the Aurora was bright enough to show them. Of thirty-four Auroræ examined by Mr. Backhouse, fourteen showed the lines 4640 and 4310, and three others at least one of these, while eight showed the red line. (Ångström only once saw this line.) In five Auroræ, all more or less red, he saw a faint band, the wave-length of which he placed at 5000 or 5100. He never saw the line 5320 (also Winlock's coronal line), unless it were once, probably from want of instrumental power. With regard to these observations, I may say that with a Browning's miniature spectroscope I saw only two lines (the red and the green) in the grand display of the 24th October, 1870; and with an instrument of larger aperture the green line only on the 4th February, 1872; while I saw the green line and three others towards the violet with the same instrument during the Aurora of 4th February, 1874. (See description of this Aurora, *antè* p. 21, and drawing of spectrum, Plate VI. fig. 1 *a*.)

Lord Lindsay's Aurora-Spectrum, 21st October, 1870.

Lord Lindsay's Aurora of 21st Oct., 1870.

Lord Lindsay observed a fine Aurora at the Observatory at Dun Echt on the night of the 21st October, 1870. It commenced about 9.30, reached its maximum about 11, and faded away suddenly about 11.30 P.M.

Spectrum described.

A spectrum obtained in the north-west gave five bright lines with a Browning's direct-vision spectroscope—two strong, one medium, two very faint. A tallow candle was used to obtain a comparison spectrum of sodium and carburetted hydrogen.

A drawing of the spectrum obtained is given on Plate XI. fig. 2. No. 1 is a sharp well-formed line visible with a narrow slit.

No. 2, a line very slightly more refrangible than F. The side towards D is sharp and well defined, while on the other side it is nebulous.

No. 3, slightly less refrangible than G, is a broad ill-defined band, seen only with a wide slit.

No. 4, a line near E, woolly at the edges, but rather sharp in the centre. This, says Lord Lindsay, should be at or near the position of the line 1474 of the solar corona.

No. 5, a faint band, coincident with b, extending equally on both sides of it.

The lines are numbered in order of intensity. It is questionable, from observations with instruments carrying a scale, whether the line-positions are exact; but the description of their characters is valuable.

Candle-spectrum.

As a candle blue-base spectrum is at times a ready and handy mode of reference in Auroral observations (as was found in this instance), I have, on Plate XI. fig. 5, given a representation of it as seen with my Auroral spectroscope. Dr. Watts's corresponding carbon-spectrum is added on the lower margin. The numbers on the upper margin refer to my scale.

Spectrum of the Aurora Australis.

Captain Maclear's spectra of Aurora Australis.

Captain Maclear, on examining the streamer seen by him Feb. 9th, 1874 (*antè*, p. 27), with the spectroscope, found three prominent lines in the yellow-green, green, and blue or purple, but not the red line. In the Aurora of March 3rd, 1874 (p. 27), he could trace four lines, three bright and one rather faint. They must have been exceedingly bright to show so plainly in full-moon light.

Instrument used, and mode of registering lines.

The spectroscope used was a Grubb single-prism with long collimator. A needle-point in the eyepiece marked the position of the lines; and a corresponding needle-point, carried on a frame by a screw movement in concord with the point of the eyepiece, scratched the lines on a plate of blackened glass. Two plates were taken. On the first were scratched the auroral lines and the solar lines as seen in the moonlight; on the second plate were scratched the auroral lines, the Solar lines from the moon, and the carbon lines in a spirit-lamp.

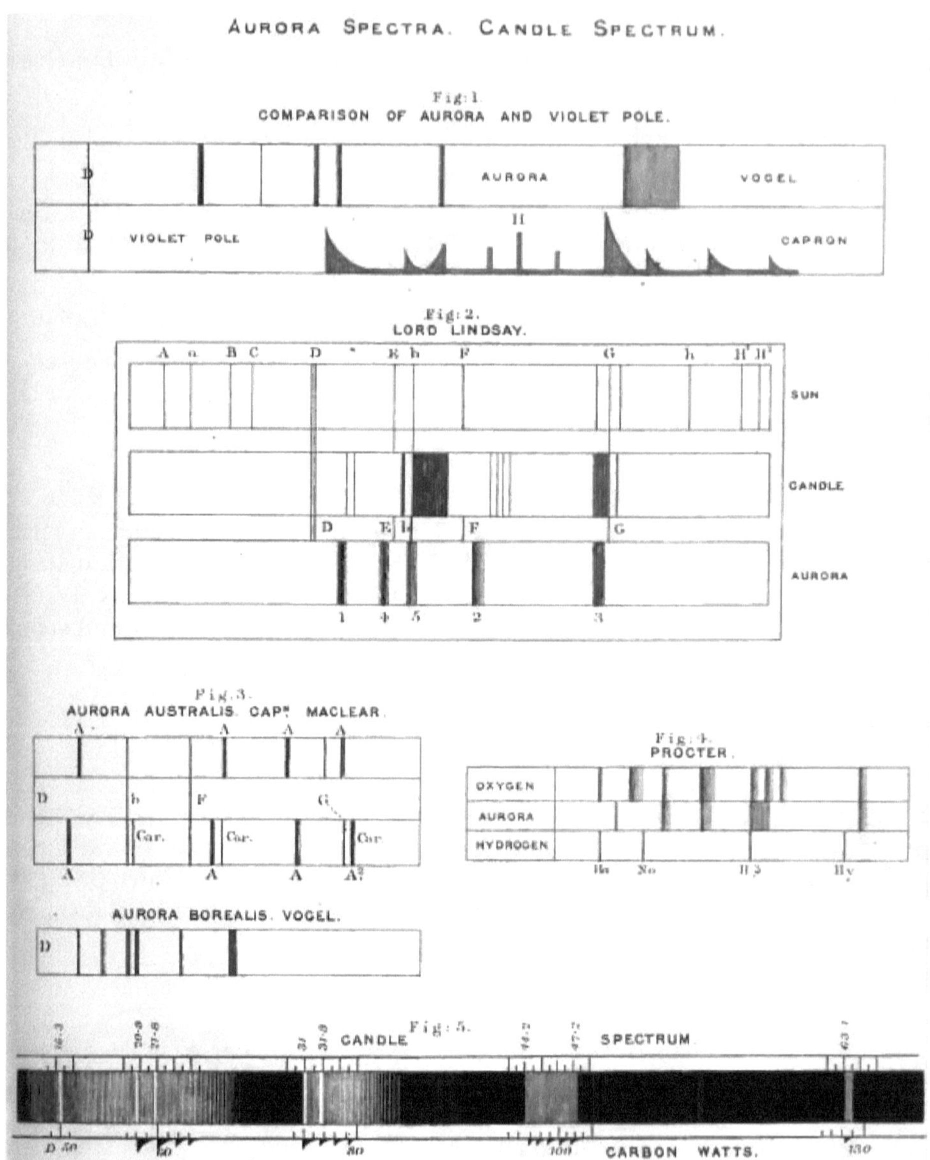

Plate XI.

AURORA-SPECTRA, CANDLE-SPECTRUM

Copies of the two spectra obtained. Discrepancy in the spectra. Remarks on the spectra.

The next morning the solar lines were verified in sunlight. I subjoin (Plate XI. fig. 3) copies of the two spectra as printed in 'Nature,' the auroral lines being marked A, the solar lines by the usual designating letters, and the carbon by *Car.* To these spectra I have added for comparison Dr. Vogel's spectrum of the Aurora

Borealis. Captain Maclear could not account for the different positions of the auroral lines in the two plates; for the prism, as far as he was aware, was not moved during the observations. As the solar lines are indicated in the same place in both spectra, the case would seem one of actual change of position of the auroral lines during observation. A comparison of the two spectra gives the impression that the lower one is the same as the upper, except that the dispersion is greater, the lines remaining relatively in position. One does not, however, see how the dispersion could have so varied in a single-prism instrument, and the position of the solar lines is adverse to such an explanation.

There is a suspicion that Auroræ are not always identical in position of some of the lines; but the line in the green (considerably out of place in the lower Australis spectrum) has always, within small limits, the same position. It will be noticed how much further the Australis spectrum runs into the violet than Vogel's Borealis, the latter having no lines much beyond F.

The faint line (No. 2) mentioned by Captain Maclear possibly corresponds with Dr. Vogel's band. The absence of the four lines of the Aurora Borealis in the green part of the spectrum of the Australis is peculiar; and in this respect, too, the two Australis spectra agree.

Comparison of the lines.

The nearest approaches to Captain Maclear's lines (of the upper spectrum) which I can find are:—

Line	Corresponding line.
1.	5567, Ångström.
2.	(The faint line.) Vogel's band, 4694-4629.
3.	I find no approximately corresponding line.
4.	4350, Alvan Clarke.

But the comparisons are not by any means close. Further observations of the Australis spectrum are very desirable.

Prof. Piazzi Smyth's Aurora-Spectra.

Prof. Piazzi Smyth's chemical and auroral spectra.

Prof. Piazzi Smyth, in volume xiv. of the 'Edinburgh Astronomical Observations,' 1870-77, has compared simultaneously the Aurora-spectrum with the sets of bright lines seen in the blue base of flame—the lines of potassium, lithium, sodium, thallium, and indium being also introduced for comparison. The spectra are drawn as seen under small dispersion, and will prove most useful in cases where an Aurora is not bright enough to admit of the lines being measured by micrometer, and the eye and comparison spectrum are obliged to be resorted to.

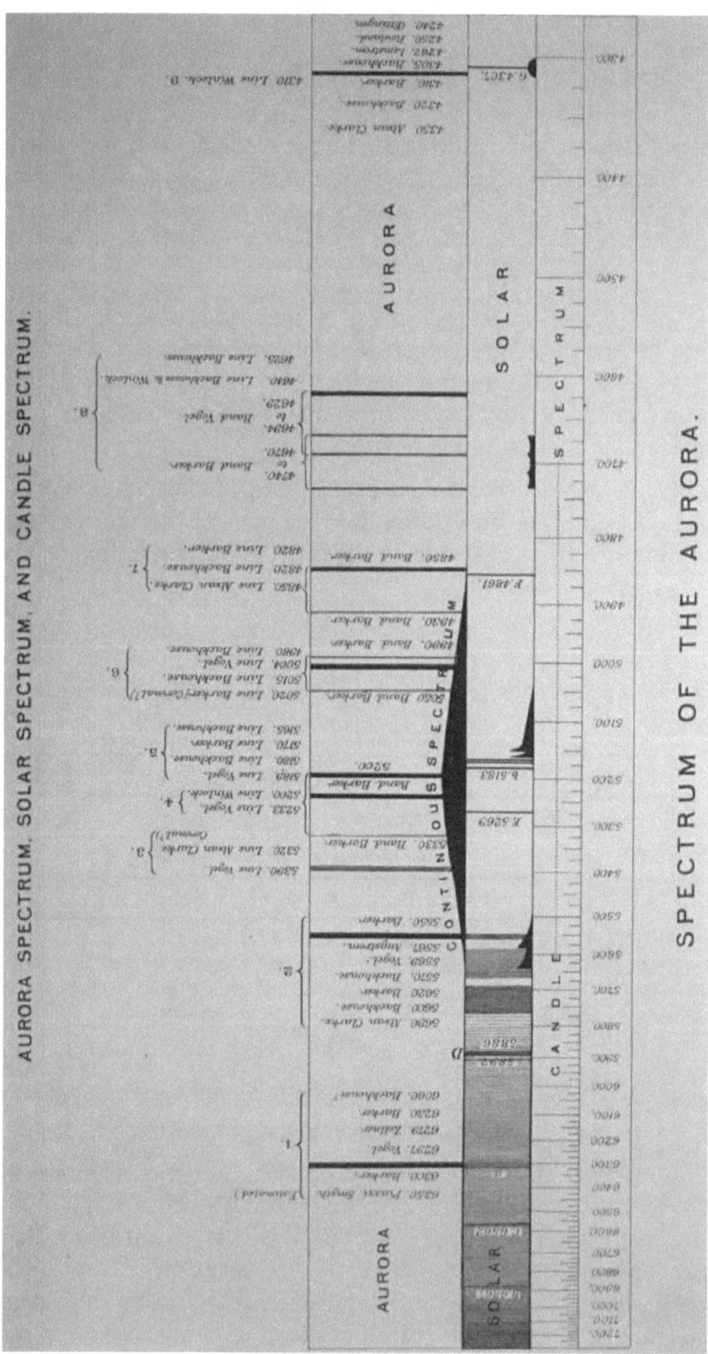

Plate XII.

AURORA-SPECTRUM, SOLAR SPECTRUM, AND CANDLE-SPEC-
TRUM

Author's Catalogue of the Auroral Lines.

(See Plate XII.)

1. W.L. 6297, Vogel. Very bright stripe; first noticed by Zöllner. Seen only in red Auroræ; stands out on a dark ground, without other lines near it. Character of line sharp and well defined; varies in colour from dusky red to bright crimson. Intensity, Herschel, 0 to 4 or 8. According to same, position coincident with atmospheric absorption-group "*a*" in solar spectrum (between C and D). I confirm this position according to my scale of solar lines, and a drawing of the coincidence (in which, and in Plate XII., the absorption-lines are drawn too dark) is given on Plate XIII. fig. 2. Herschel says this line coincides with a red band in the negative glow-discharge, but its identity is doubtful. Its isolation and want of adjacent lines seem to separate it from the air-spectrum and gas-spectra in general. At the appearance of this line, 5569 (No. 2) becomes noticeably fainter. When this line is bright, 5189 (No. 5) is bright also (Vogel).

If we propose to assign to this line, as well as to 5569, a phosphorescent origin, it would be strongly confirmatory of such a theory (in connexion with the phosphoretted-hydrogen spectrum) to find it brighten at low temperatures.

Note.—Sir John Franklin says, in his 'Polar Expeditions,' that a low state of temperature is favourable for the production of brilliant coruscations. It was seldom witnessed that the Auroræ were much agitated, or that the prismatic tints were very apparent, when the temperature was above zero.

2. Line in the yellow-green. Brightest of all lines in the Aurora-spectrum. W.L. 5567, Ångström; 5569, Vogel. Intensity 25, Herschel. To me more pale green than yellow, sometimes flickering and changing in brightness (Herschel and Capron). Seen in all Auroræ usually sharp and bright, but Procter has once recorded it nebulous. Its character as to width, sharpness, and intensity, if carefully observed, might indicate height and structure of Aurora. Becomes noticeably fainter at appearance of red line (Vogel). Found by me to correspond in position with a faint atmospheric absorption-band (see Plate XIII. fig. 2). According to Ångström and Herschel, arising from a phosphorescent and fluorescent light, emitted when air is subjected to the action of electrical discharge.

3. Line in green near last. W.L. 5390. An extremely faint and unreliable observation (Vogel). Seen only by him, unless Alvan Clarke's 5320 (coronal?) be the same.

4. Line in green-blue. W.L. 5233, moderately bright (Vogel); 5200, Winlock. Intensity, 2 or 0? to 6, Herschel. Coincides with line in the negative glow according to same. Frequently observed.

5. Line in green-blue. W.L. 5189. This line is very bright when the red line appears at the same time; otherwise equal in brilliancy with No. 3 (Vogel); Winlock, 5200. Not so frequently observed as No. 3. Barker gives a band extending from 5330 to 5200. Intensity of 5189, 0 to 8, Herschel, who considers it coincident with a constant strong line in the spark-discharge.

6. Line in blue. W.L. 5004. Very bright line, Vogel; 5020, Barker (coronal?).

Intensity 2 or 0? to 8, Herschel. Coincides with line of nitrogen in the nebulæ according to same. Barker gives a band extending from 5050 to 4990.

7. Line in the blue not found by Vogel in Aurora, April 9th, 1871. W.L. 4850, Alvan Clarke; 4820, Backhouse and Barker. Intensity, Herschel, of 4820-4870, 0 to 4? Herschel suspects this and No. 4 to be seen only in Auroral streamers of low elevation. Barker gives a band extending from 4930 to 4850.

8. 4694, 4663, 4629. Broad band of light, somewhat less bright in the middle; very faint in those parts of the Aurora in which the red line appears (Vogel). Intensity 3-6 (Herschel). A double band, consisting of two lines, the first rather more frequently noted than the second in Auroral spectra, agrees well in position with the principal band in the negative glow-spectrum (same). Barker gives a band extending from 4740 to 4670; Backhouse and Winlock give a line at 4640, situate within the same.

9. There seems a good deal of confusion about a fairly bright line (intensity 0-6, Herschel) seen in most Aurorae (not, however, by Vogel, April 9th, 1871), and situate somewhere near G in the solar spectrum. Alvan Clarke places it at 4350, on the less refrangible side of G; Backhouse and Barker at or very near to G; while Lemström and others position it on the more refrangible side of G. Accurate observations, for which a quartz spectroscope might be useful, are much wanted. Herschel makes this line, at 4285, correspond with a strong band in the violet in the negative glow-spectrum.

Herschel also refers to an apparently additional line near the hydrogen-line, or between G and H_1, in the solar spectrum, as mentioned once by Lemström at Helsingfors. I am not aware of any other observation of this line, which must be considerably beyond that at or near G, and would probably be difficult to detect, except in instruments specially adapted for examination of the violet end of the spectrum.

Theories in relation to the Aurora and its Spectrum.

Lemström's.

Lemström (1):—That the Polar light is caused by an electric current passing from the upper rarefied layers of the air to the earth, producing light-phenomena that do not arise in the denser layers of the air. (2) That there are nine rays (lines or bands) in the Aurora-spectrum, which in all probability agree with lines which belong to the gases of the air. (3) That the Aurora-spectrum can be referred to three distinct types, which depend on the character of the discharge.

Vogel's.

Dr. Vogel:—(1) That the Aurorae are electric discharges in rarefied-air strata of very considerable thickness. (2) That the Aurora-spectrum is a modification of the air-spectrum, involving the question of alteration of the spectrum by conditions of temperature and pressure.

Ångström's.

Ångström:—(1) seems to adopt the hypothesis that the Aurora has its final cause in electrical discharges in the upper strata of the atmosphere, and that these, whether disruptional or continuous, take place sometimes on the outer boundary of the atmosphere, and sometimes near the surface of the earth.

(2) That the Aurora has two different spectra.

(3) That the green line is due to fluorescence or phosphorescence, and that there is no need to resort to Dr. Vogel's variability of gas-spectra according to circumstances of pressure and temperature.

(4) That an agreement exists between the lines of the Aurora (except the red and green before mentioned) and the lines or bands of the violet light which proceed from the negative pole in dry air.

Zöllner's remark as to temperature of Aurora and character of spectrum.

Zöllner has pointed out that the temperature of the incandescent gas of the Aurora must be exceedingly low, comparatively, and concludes that the spectrum does not correspond with any known spectrum of the atmospheric gases—only because, though a spectrum of our atmosphere, it is one of another order, and one which we cannot produce artificially.

CHAPTER XI.

THE COMPARISON OF SOME TUBE AND OTHER SPECTRA WITH THE SPECTRUM OF THE AURORA.

[In part from an Article in the 'Philosophical Magazine' for April 1875.]

Testing Ångström's Aurora theory. Battery and spectroscope described. Vogel's spectrum selected for comparison.

In order to test Professor Ångström's theory of the Aurora, referred to in the last Chapter, in an experimental way, I examined, in the winter of 1874, some tube and other spectra, not only for line-positions, but also for general resemblance to an Aurora-spectrum. It did not seem desirable to use powerful currents. A ½-inch-spark coil, worked by a quart bichromate-cell, was found sufficient to illuminate the tubes steadily. The spectroscope used was one made for me by Mr. Browning specifically for Auroral purposes, and of the direct-vision form, being the same instrument as is described *antè*, p. 91, and figured in Plate X. fig. 1. The micrometer was the diaphragm one, also before described and figured on same Plate, figs. 2, 3, and 4. I selected Dr. Vogel's spectrum for comparison, it being, so far as I am aware, the most accurately mapped, with regard to wave-length, at one observation, of any Auroral spectrum. It seemed an unsafe plan to attempt to obtain an average Aurora by comparison of different observations made at various times by different observers with all sorts of instruments—the difficulty, too, being increased by the suspicion that the spectrum itself at times varies in number and position as well as intensity of its lines.

Central part only of spectrum mapped.

In most cases the central part of the spectrum only (corresponding to the central lines of the Aurora) was mapped, the red line in the Aurora not being found to correspond with any prominent line in the gas-spectra examined, and the Auroral line near solar G being so indefinitely fixed as to render comparison almost valueless. (See Plate XIII. fig. 1.)

Dr. Vogel's spectrum does not comprise the line near G; but I have added this (in an approximate place only) in order to complete the set of lines. For drawing of Dr. Vogel's spectrum, with its scales attached, see Plate XIII.

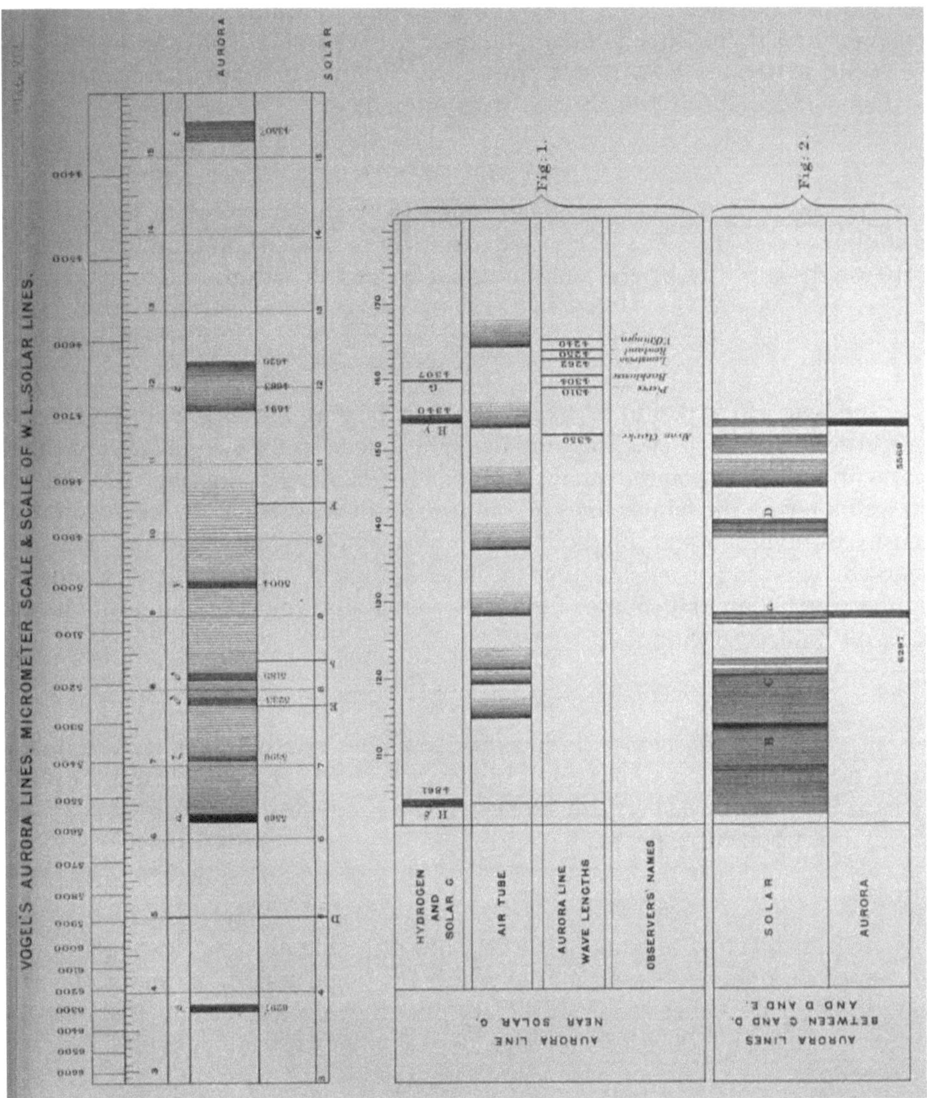

Plate XIII.

VOGEL'S AURORA-LINES, AURORA-LINES NEAR G, AND IN THE RED AND GREEN

Hydrogen-tube.

Hydrogen-tube. Colour of glow varied with intensity of current.

This tube was one of Geissler's and of rather small calibre. On illumination the wide ends were easily lighted with a silver-grey glow, having a considerable amount of stratification. The capillary part glowed brilliantly with silver-white, bright green, and crimson light, according to the intensity of the current. With the commutator slowly working, white running into green and bright green were the main features of the thread of light; on the current passing more rapidly, the capillary thread became of an intense crimson, at the same time apparently increasing in diameter (an effect probably due to irradiation).

Spectrum described.

The spectrum was very brilliant, consisting of the three bright lines usually distinguished as Hα, Hβ, and Hγ, and a number of shaded bands and fainter lines between these, with a bright continuous spectrum as a background to the whole.

Lines α, β, and γ varied in intensity with colour as seen by eye. Fainter lines or bands described.

The lines Hα, Hβ, and Hγ were found to vary in intensity with the current, and in accordance with the colour of the light as seen by the eye—a fact, as I think, not without bearing on the question of the Aurora, the varying tints of which are so well known. The fainter lines or bands were mostly stripes of pretty equal intensity throughout, and all about the width of the Hβ line. I did not trace any marked degrading on either side of the lines, though the edges were not uniformly so sharp as Hα and Hβ. Some of the lines were found coincident in position with lines of the air-spectrum.

Purity of subsidiary lines questioned.

It is a question whether these subsidiary lines are hydrogen, or are due to some tube impurity. A photograph I have taken of this tube-spectrum shows 17 lines in the part of the spectrum between F and H_2, some of which are repeated in the hydrocarbon-tube spectra.

Coincidence of lines with Aurora-spectrum.

No principal line, and one subsidiary line only, actually coincide with the Aurora-spectrum, this last being that to which Dr. Vogel assigns an identical wave-length, viz. 5189. Other of the subsidiary lines, however, fall somewhat near the Aurora-lines 5569, 5390, 5233, and 5004, two faint lines also falling within the band 4694 to 4629.

Comparison of the lines.

The lines (adopting Dr. Vogel's wave-lengths for the H lines) were, when compared, as under:—

Aurora	5569	5390	5233	5189	5004	4694 to 4629	band.
Hydrogen	5555	5422	...	5189	5008	4632	

I remarked that a line (5596) described by Dr. Vogel as "very bright" in his H spectrum does not appear in my tube, though in most other respects our H spectra agree.

Effect of distance on the spectrum.

I thought this tube afforded a good opportunity for testing the effect of distance upon the spectrum. The slit was made rather fine. At 6 inches distance from it the line α in the blue-green (F solar) was very bright. The lines marked β, γ, δ, ε, and ζ also survived, but were faint. At 12 inches from the slit α and γ were alone seen, and at 24 inches α stood by itself upon a dark ground. I also noticed that the red and yellow parts of the spectrum first lost their light on the tube being withdrawn from the slit; and this appeared to account for β disappearing while γ survived. For drawing of the hydrogen-tube spectrum see Plate XIV. spectrum 1.

The question of effect of distance upon the spectroscopic appearance of a glowing light, as tested for this and other tubes, seems an important one. It may possibly account for the generally faint aspect of the lines in the more refrangible part of the Auroral spectrum.

Carbon- and Oxygen-tubes.

Carbon- and oxygen-tubes. Tubes described. Carbon-tubes lighted up. Spectra of the carbon-tubes described.

The following tube-observations were taken together, because my friend Mr. Henry R. Procter (to whom I am indebted for many profitable hints and suggestions in Auroral work) was disposed to regard the spectra found in the carbon-tubes and in those marked "O" as identical, suggesting that pure O, with the ordinary non-intensified discharge, gives only a continuous spectrum; and that the O tubes are in fact generally lighted up by a carbon-spectrum, as the result of impurity from accidental causes. The tubes examined for the purpose of comparison were as follows:—A coal-gas tube, a tube marked "C.A.," three O tubes (two of, I believe, London make, and the third from Geissler), and an OH_2 tube, also from Geissler. The carbon-tubes were both brilliantly and steadily lighted by the current. The C.A. tube glowed with a peculiar silvery-grey green light in the capillary part, and with a grey glow, considerably stratified, in the bulbs. The coal-gas tube illumination was whiter and still more brilliant than the C.A., and with even finer stratification in the bulbs. The spectra of both tubes were conspicuous for the same three well-known principal bright lines or bands in the yellow, green, and blue (with one fainter in the violet), all shading off towards the violet, and in both cases with fainter intervening bands or lines. These last bands or lines only partially coincided when the two tubes were compared.

The spectra in both cases were rich and glowing, with a certain amount of continuous spectrum between the lines; and the three principal bands or lines showed well and distinctly their respective place-colours.

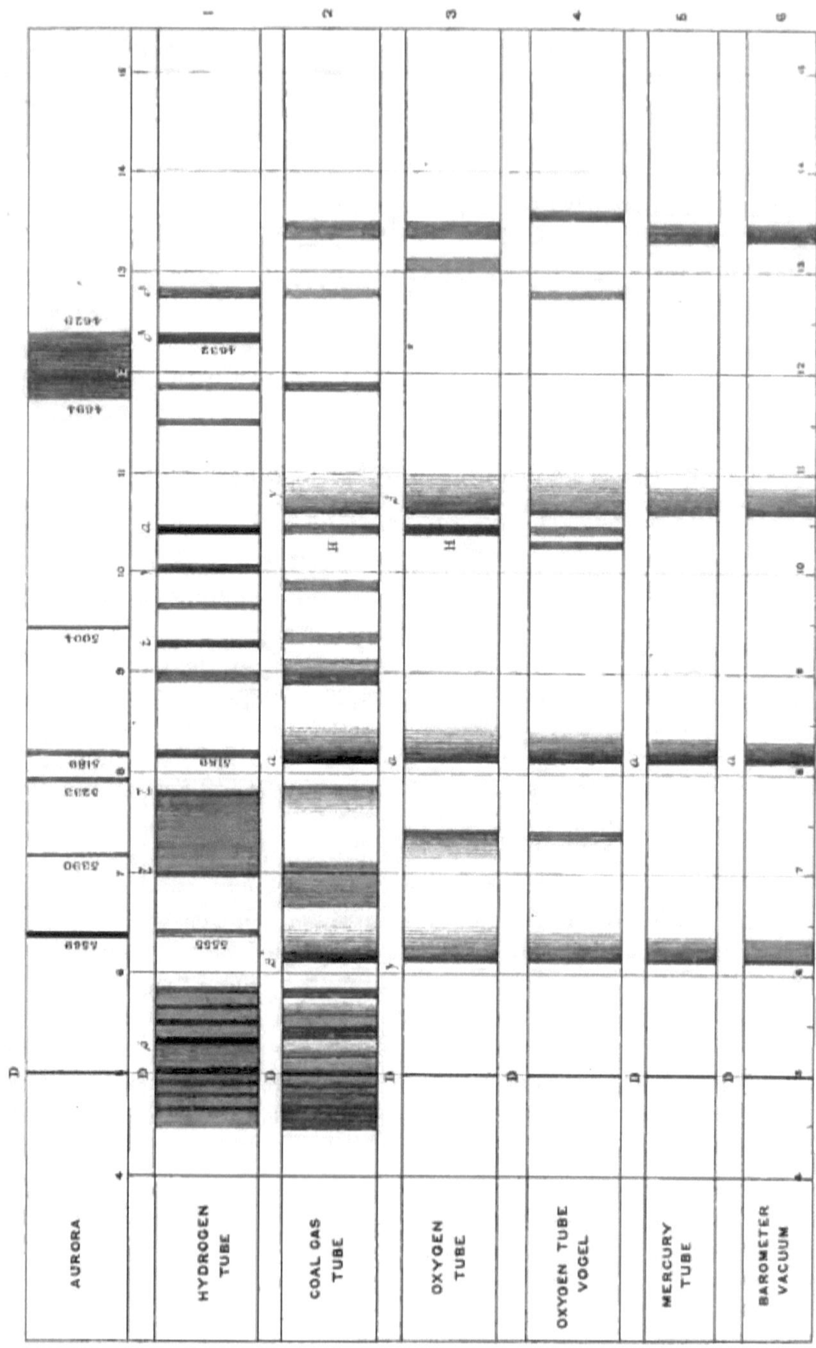

Plate XIV.

AURORA, HYDROCARBONS, OXYGEN

Tubes tested for distance.

Tubes tested for distance. —In the case of the C.A. tube, at 18 inches from the slit the continuous spectrum and fainter lines disappeared, while the four principal lines still shone out, that in the green being the strongest. At 24 inches the same lines were still visible, though somewhat faintly.

In the case of the coal-gas tube, at 24 inches the whole spectrum was quite brilliant, the four principal lines being very bright and even preserving their distinctive colours. The H line, near the line or band in the blue, was also plainly seen.

O tubes lighted up.

The O tubes, when treated by the same current as the carbon-tubes, were found to be all three identical in general features. The discharge lighted up each of the tubes feebly and somewhat intermittently. Grey in the bulbs and a faint but decidedly pinkish white in the capillary part were the distinguishing light colours; while nothing could be more marked than the difference in brilliancy between these and the preceding carbon-tubes.

OH$_2$ tubes lighted up. O tubes spectra described.

The OH$_2$ tube presented very much the same character; but the discharge occasionally varied from a pinkish white to a yellow colour, somewhat like that which artists call brown-pink, and reminding one of the "golden rays" in certain Auroræ. These O spectra presented, in common with the carbon-tubes, three principal bright lines or bands in the yellow, green, and blue, with a fainter one in the violet, all shading off towards the violet. The bands, however, showed but very little trace of local colour, and the whole spectrum had a faint and washed-out look, very different from the carbon-spectra. (I certainly, by a little management, subsequently succeeded in getting the same look to the C.A. spectrum; but it was only by removing the tube to some distance from the slit, and thus depriving the spectrum of very much of its brightness.) The hydrogen line (solar F) was bright, more so than any of the O lines.

The intensity of the three principal lines seemed to me to run in the following order:—

	Yellow.	**Green.**	**Blue.**
Coal-gas	β	α	γ
Oxygen	γ	α	β

Between the lines γ and α in the Geissler O tube I found a rather bright line, which I shall have occasion to refer to hereafter.

At 12 inches distance from the slit the O spectrum lost nearly all its light; the H line, and the three lines γ, α, and β, alone faintly remaining, α being decidedly the brightest. At 24 inches no spectrum at all was to be seen.

Comparison of spectra of coal-gas and O tubes.

I carefully compared together the three principal lines of the two spectra of coal-gas and O by means of:—

1st, the photographed micrometer before described;

2nd, a comparison-prism on the slit plate;

3rd, a piece of very fine brass foil cut as a pointer and fixed in the focus of a positive eyepiece.

The lines or bands in both tubes were found to be slightly nebulous towards the less-refrangible end (where they were measured), and the O tube was not bright under a moderately high power (positive eyepiece). Subject to these remarks, the three principal lines in both tubes were found to correspond in position within the limits of my instrument. The spectra did not, however, I am bound to say, *look* alike.

Dr. Vogel's O spectrum reduced and compared.

Puzzled by these observations, it then occurred to me to reduce Dr. Vogel's spectrum of O, given in his memoir, to the same scale with my own. This I did independently, and I then compared the result with my own spectrum as mapped out. From the comparison, I judge that if my O tubes, one and all, showed a carbon-spectrum, the learned Doctor's tube must have been subject to a similar infirmity, as the tubes all agreed in main features.

There is, however, one point to which I desire to draw attention, which is this, that common to both the Doctor's and my own Geissler spectrum I found the before-mentioned rather bright line between γ and α. This line I found no equivalent for in either of the carbon-tubes. For spectra of coal-gas and oxygen-tubes, see Plate XIV. spectra 2, 3, & 4.

Tube- and flame-spectra of carbon do not correspond.

In comparing the spectra, it should be remembered that the tube- and flame-spectra of carbon do not correspond. Compare, for instance, the spectrum of coal-gas or CO_2 in tube, and the well-known lines or bands in the blue base of a candle-flame. The sharper edge of the yellow line or band of the carbon-tubes will be found about midway between the two bright yellow candle-lines or bands. The first of the very beautiful group of lines or bands in the green in the candle-flame falls considerably behind the sharper edge of the green line or band in the tube, while the third bright band in the tube, alone of the three, corresponds with a very faint band in the candle-flame. A line or band in the violet in the tube-spectrum finds no equivalent in the candle-spectrum. For comparison of the carbon-tube and flame spectra (the principal lines of the tube being alone shown), see Plate XVI. spectra 6 & 7.

Prof. Piazzi Smyth's measurements of the components of the citron-band in a coal-gas flame.

Note.—Prof. Piazzi Smyth has been good enough, at my instance, to measure the components of the citron band of the carbo-hydrogen spectrum (near Ångström's Aurora-line), as seen in a coal-gas blowpipe-flame urged with common air.

The spectroscope used had prisms giving 22° of dispersion between A and H, and the observing telescope magnified 10 times. The following is a table of the results communicated to me by the Professor:—

	Intensity.	Reading of Micrometer.
Reference line, lithium β	4	16·55
Reference line, sodium, α1	10	18·45
Reference line, sodium, α2	10	18·51
Citron band. Carbo-hydrogen.		
Line 1, exquisitely clear	6	21·28
2, exquisitely clear	5	21·88
3, exquisitely clear	3	22·44
4, faint but clear	2	22·95
5, faint	1	23·38
6, faint and hazy	1	23·70
7, doubtful	?	23·92
Reference line, thallium α	10	25·08

Comparison of Dr. Vogel's O lines and Dr. Watts's carbon-lines.

From Dr. Watts's 'Index of Spectra' I have extracted the three principal carbon-tube bands or lines; and they compare with Dr. Vogel's oxygen-tube as under:—

	Yellow.	Green.	Blue.
Dr. Vogel's oxygen-lines	5603	5189	4829
Dr. Watts's carbon-tube bands or lines	5602	5195	4834

Now these wave-length differences are so small that they raise a presumption of the possibility of the spectra being identical. On the other hand, assuming the spectra are not identical, the comparison tells the other way, viz. that the differences are so minute as to escape detection in instruments of moderate dispersion. With my own instrument I found the O spectrum too faint to increase the dispersive power with advantage. Considering the extremely different character of the two discharges, the identity of all the O tubes, and the presence of the line found between γ and α in the O spectrum, I think the two spectra are independent, though I admit there is room for doubt.

O and CO$_2$ spectra photographed.

Note.—Since this examination I have photographed both spectra side by side (see 'Photographed Spectra,' Plate XXXI., text, pp. 69, 70). The pictures include, of course, only the blue and violet parts of the spectrum; but they are widely different in aspect, and show that, photographically at least, in this part of the spectrum there is a complete want of identity. Subsequent investigations, however, by Schuster and others (detailed later in this Chapter), go to establish that the principal lines shown in mine and Dr. Vogel's tubes were due to (probably hydrocarbon) impurity. The exception is the single line common to mine and Dr. Vogel's tubes, but absent from the coal-gas spectrum. This line proves to be oxygen. Compare oxygen-tube spectra (Plate XIV. spectra 3 and 4) with Schuster's oxygen-tube spectrum (Plate XVIII. fig. 15). The line in question is found identical in the three tubes.

The tube OH_2 was found to give the principal lines of the O and H spectra combined on a faint continuous spectrum.

Geissler Mercury-tube (Plate X. fig. 7) and Barometer Mercurial vacuum.

Mercury- and barometer-tubes examined. Mercury-tube described. Barometer-tube.

I next examined two vacuum-tubes of an entirely different character. The one was a tube from Geissler of stout glass, some fifteen inches long, without electrodes, and an inch across. Within this tube was a second of uranium glass, with bulbs blown in it. In contact with both tubes a quantity of fluid mercury ran loose (Plate X. fig. 7). Upon shaking this tube with the hand brilliant flashes of blue-white light, like summer lightning, flashed out. These were discernible (though faintly) even in daylight. The fine terminal wires of the coil being wrapped round each end of this tube, when the current passed, a bright and white induced discharge, with a considerable amount of stratification, was seen in the tube. The other tube was that of a mercurial siphon-barometer. This being placed in a stand, one terminal wire was placed in the mercury in the short leg of the siphon, while the other terminal was made into a little coil and placed on the upper closed extremity of the barometer-tube. On passing the current, the entire short space above the mercury was filled with a grey-white light, not stratified, but showing a conspicuous bright ring just above the level of the mercury.

Spectrum of both these tubes described.

Both these tubes, when examined with the spectroscope, showed four bright rather uniform bands (the central one being the brightest), which I assigned to the carbon-spectra (see Plate XIV. spectra 5 and 6).

The Geissler tube was probably filled designedly with coal-gas. In the case of the barometer-tube the spectrum must be assumed to be the result of some carbon impurity.

No lines of mercury could be detected in either case.

An effort was made to examine the light of the Geissler mercury-tube as excited by motion only, but the spectrum could not be kept in the field; the four lines were, however, seen to flash out as the light passed before the slit.

Air-tubes.

Air-tube illuminated.

The first tube I examined was an ordinary Geissler tube charged with rarefied air. The bulbs, on passing the discharge, were filled with the well-known rose-tinged light like to the Aurora-streams. This in the capillary part was condensed into a brighter and whiter thread, while the platinum wire of the negative pole was surrounded by its characteristic mauve or violet glow.

Spectrum described.

The spectrum, even with a weak current, was quite bright, and consisted mainly of the nitrogen-lines and bands, with the lines Hα, Hβ, and Hγ, and some of the intermediate lines of the H tube.

The double line α was undoubtedly the brightest in the spectrum when taken in the capillary part of the tube. After this followed β, and then γ(H), δ, and ε. I was, however, uncertain as to the relative brightness of the last three, and marked their intensities with hesitation. I tested them several times independently with differing results, and suspected them of variability with the current.

The rest of the lines were very much of the same intensity. (For drawing of spectrum of air-tube in capillary part see Plate XV. spectrum 1.)

Violet [negative] Pole, same tube.

Violet (negative) pole: spectrum described.

I next turned my attention to the violet or negative-pole glow; and here a remarkable change took place in the spectrum, not only in the position of the principal bands or lines, but in their relative intensity (see Plate XV. spectrum 2).

The double line α in the capillary part was replaced in the violet glow by a shaded band of second intensity β, the sharp edge of which was extended towards the red, and formed (except for some faint indications) the limit of the spectrum in that direction. The somewhat faint line next α in the capillary tube had its faint representative in the violet pole; but the next two lines (capillary) were represented by the bright band γ in the violet pole lying in a position between them. Next γ in the violet pole came three faint lines, representing β, γ, and δ in the capillary spectrum; and then the bright band α, which was the brightest of the violet-pole group, and represented a medium-intensity band in the capillary spectrum. After this was a faint band near α, representing two rather bright ones in the capillary spectrum, this last being succeeded by other bands in the violet. α, β, and γ in the violet pole were examined carefully for relative brightness, and were, I believe, correctly marked.

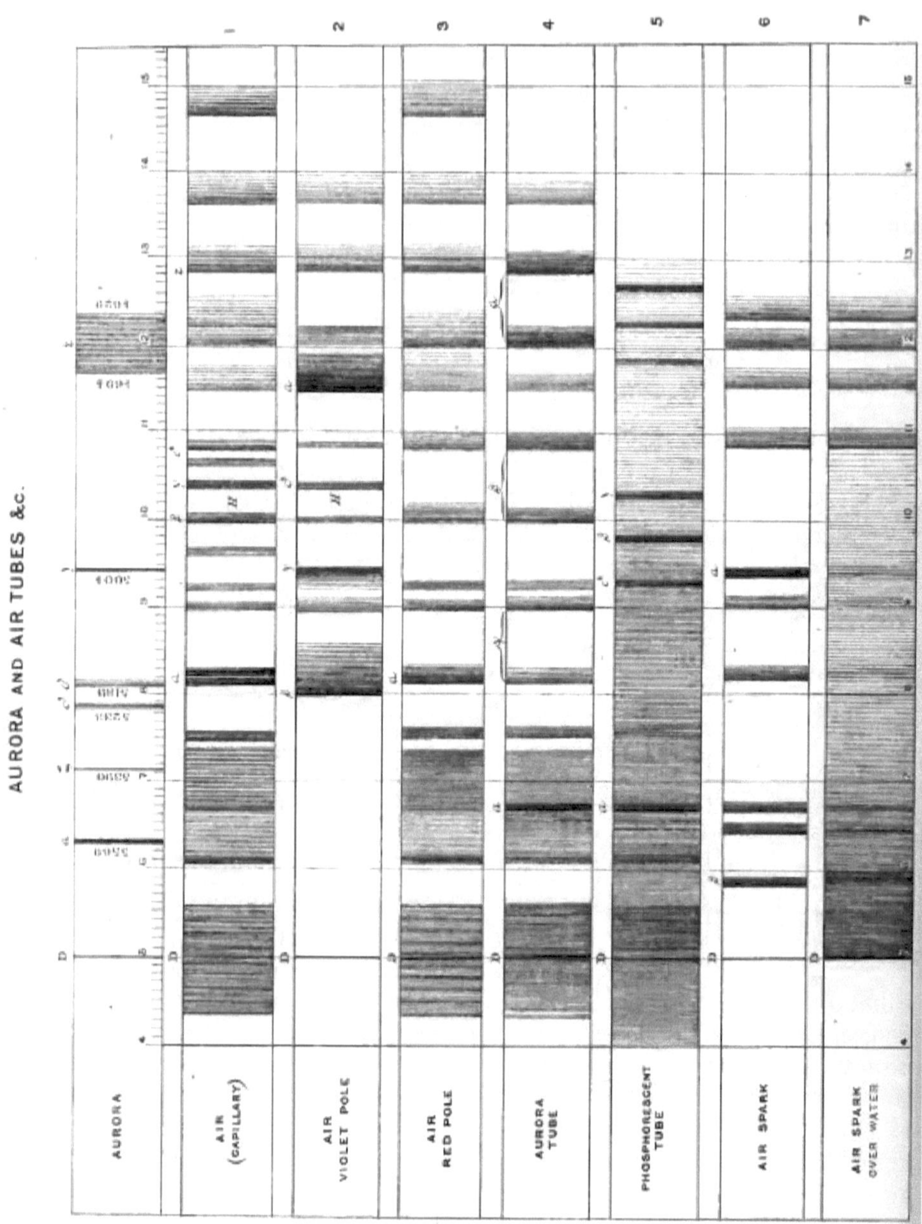

Plate XV.

AURORA AND AIR-TUBES, &C.

Red [positive] Pole.

Red (positive) pole: spectrum described.

The red [positive] pole was next examined, but presented no peculiar features. It appeared as a fainter representation of the capillary air-spectrum, with some few lines or bands absent, and (as will be seen after) was also a fair representation of a diffused air-spectrum (see Plate XV. spectrum 3).

Examined for comparative intensity, at 24 inches from the slit, the whole capillary air-spectrum showed faintly. The marked lines in the centre of the spectrum generally retained their prominence; but after α I judged ε next in brightness. On examining the violet pole at 12 inches from the slit, the whole spectrum was faint and the bands α and β were alone distinctly seen.

Aurora (air)-tube. (Plate XV. spectrum 4.)

Aurora-tube: discharge described. Spectrum described.

Next to the Geissler air-tube I examined an "aurora"-tube, about 15 inches long and 1¼ inch across, with platinum terminals, and of the same diameter throughout (Plate X. fig. 8). The discharge was of a rosy-red colour, and the long flickering stream from pole to pole certainly much reminded one optically of an auroral streamer. Spectroscopically examined, the discharge presented a faint banded air-spectrum similar to that of the positive pole (see Plate XV. spectrum 4); but the relative intensity of the lines was somewhat altered, while a very bright line in the green (seen also in the tube next described) was characteristic of the spectrum, and in this respect distinguished it from the ordinary air-spectrum.

Phosphorescent tube.

Phosphorescent tube described. Discharge described. Spectrum described.

Following this last tube I examined one purchased as "phosphorescent." It was rather short (6½ inches), of equal calibre, and about the size of the bulb of a Geissler tube. It was filled with a white powder (probably one of the Becquerel compounds). On passing the current between the electrodes, a bright rose-coloured stream appeared; and wherever this was in contact with the powder, the tube glowed with a brilliant green light. On stopping the current, the tube still continued to shine, but with a fainter green glow, which gave only a continuous spectrum. When examined in full glow, the tube-spectrum was also in the main continuous and of a green tinge; but upon it were bright lines in the blue and violet portions of the spectrum, while in the red, yellow, and green a faint but distinct air-spectrum was seen; and with this was also found the same bright line in the green which distinguished the "aurora"-tube. [Five out of six of the lines in the blue and violet will be also found in Schuster's oxygen-tube, violet pole (Plate XVIII. fig. 15). The air-spectrum probably arose from impurity.]

Spark in Air.

Spark in air: spectrum described.

I next took a ½-inch spark in air between platinum terminals (see Plate XV. spectrum 6). The principal lines in this spectrum were the line α (by far the brightest), corresponding to γ in the violet pole; next was β, a line in the yellow, not appearing in the tube-spectrum, and then other lines of less intensity. In the "aurora" and "phosphorescent" tubes was found, as before mentioned, a line in the green prominent for its brightness, and, indeed, in the "aurora"-tube the only one which survived when it was moved away from the slit. This line also appeared in the spark-spectrum, but there only of an average brightness. I examined it carefully for position in the respective tubes; and on comparing them by means of a pointer in the eyepiece, found it coincident with the ridge or centre of the wedge-like bright-green broad band which is so conspicuous in the air-tube spectrum.

I think this edge-like centre has actually a line coincident with the line I refer to; but if so, its intensity little exceeds that of the band itself.

Spark over Water.

Spark over water: spectrum described.

To complete the set of air-experiments, I examined the same spark taken from the surface of a small meniscus of water, placed in a glass cup upon the lower platinum wire. In this case the air-spectrum was plainly, but not brightly, seen at the violet end of the spectrum—the red, yellow, green, and blue being filled with a continuous spectrum, through which some of the air-lines faintly showed (see Plate XV. spectrum 7).

Phosphoretted-Hydrogen Flame.

Phosphoretted-hydrogen flame.

This was obtained from a hydrogen-bottle fitted with glass tubing, two or three minute pieces of phosphorus being placed with the zinc. The flame was of a bright yellow colour, with a cone of vivid green light in its centre.

Spectrum described.

The spectrum was found to consist mainly of three bright bands in the yellow, green, and green-blue respectively (see Plate XVI. spectrum 3).

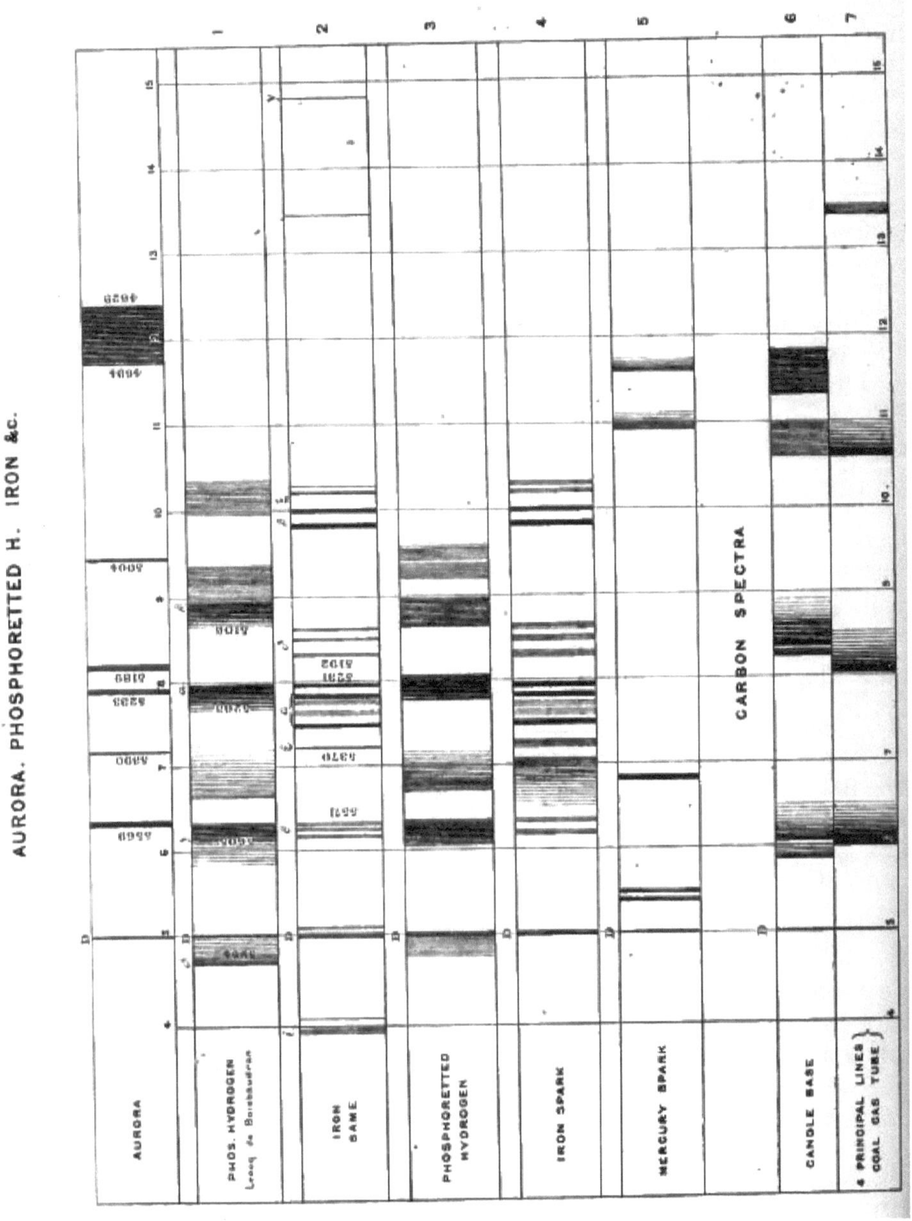

Plate XVI.

AURORA, PHOSPHORETTED HYDROGEN, IRON, &C.

Mons. Lecoq de Boisbaudran's remarks on the spectrum increasing in brilliancy when the flame is cooled.

The central band was very striking in its emerald-green colour, while all the bands were remarkable as being very broad in proportion to the slit (which, however, was not fine). The yellow band had a rich glow of colour. My spectrum was mapped out at ordinary temperature, and I found the bands sufficiently bright; but Mons. Lecoq de Boisbaudran, in his 'Spectres Lumineux' (texte, p. 188), has described how the brilliancy of these bands is increased when the flame is artificially cooled (*refroidie*).

The idea of cooling the flame was due to M. Salet, who effected it either by a jet of water or by an air-blast.

The less refrangible bands seem the most susceptible to increase of brilliancy.

Mons. Boisbaudran also makes the important remark that the relative intensities of the bands are in such case altered, adding:—"La plus importante de ces modifications consiste en un renforcement très-considérable de la bande rouge δ 97·03 (W.L. 5994) qui devient vive de presque invisible qu'elle était en l'absence du refroidissement artificiel de la flamme."

Full details of the changes are given by M. de Boisbaudran.

The bearing of these observations as connected with the variable character of the red line in the Aurora-spectrum seems to me in the highest degree noteworthy.

Iron-Spectrum.

Iron-spectrum.

A comparison of this spectrum suggested itself, partly from the suspected relations between the Aurora and solar corona, and partly from a consideration of the views expressed by M. Gronemann and others in favour of the Aurora having its origin in the fall of an incandescent meteoric powder.

How obtained. Spectrum described. Mons. Lecoq de Boisbaudran's spectra also given.

The spectrum was obtained from a spark taken over a solution of perchloride of iron in a small glass cup, and was remarkable for its brightness in and about the green region. The lines varied considerably in intensity, and with a fine slit the principal ones were sharp, distinct, and clear. A group of three lines (α) stood out boldly in the green as the most marked, and next to these a group of three others more towards the violet end of the spectrum (see Plate XVI. spectrum 4). By the side of my phosphoretted-hydrogen and iron spectra I have placed the principal lines of Mons. Lecoq de Boisbaudran's same spectra (reduced to my scale), and with figures of wave-lengths for comparison with the Aurora-spectrum (see Plate XVI. spectra 1 and 2).

Comparison of iron- and Aurora-spectrum.

A difficulty in comparing the iron-spectrum with that of the Aurora arises from the large number of fine lines found in the former spectrum. In a photograph (taken with the same prism as before described) of a small piece of meteoric iron fused in an electric arc by the aid of 40 Grove cells, about 154 lines are easily counted in the blue and violet parts of the spectrum. Double this number at least would be seen with a spectroscope of moderate dispersion in the region comprising the entire set of auroral lines.

Spectrum of Mercury.

Mercury-spectrum. How obtained.

This spectrum is given as useful for comparison with the bright and principal Aurora-line. It is easy to obtain with a small coil, the metal being used as one electrode. The yellow lines are distinct and steady; but the green, which is very bright, is apt to flicker as the spark moves on the surface of the metal (see Plate XVI. spectrum 5).

The following Table was compiled for the purpose of comparing the foregoing results with the Aurora-spectrum.

Table of coincidences.

TABLE showing comparative position of Aurora-lines with the principal lines in the examined spectra. C. means coincident within the limits of my instrument and scale, N. near, and VN. very near.

Aurora-lines	6297 β.	5569 α.	5390 ζ.	5233 δ.	5180 δ.	5004 γ.	4694 to 4629 ε.	4350? ε.
Hydro-gen-tube		N.		N.	C., same W.L.		Band includes 2 lines.	
Coal-gas tube				N.	VN.		Band includes 1 line.	
Oxy-gen-tube					VN.			
Air, capil-lary		Band in-cludes	Band in-cludes		N.		Band includes 2 lines.	
Air, vio-let-pole	No re-sults in the ex-amined spectra; but see Plate XIII. fig. 2.				Band includes	C.	Band includes 1 line.	Too un-certain in posi-tion for compar-ison (see Plate XIII. fig. 1).
Air, red-pole		See Air, capillary.						
Auro-ra-tube and phosphores-cent tube		See Air, capillary; and note bright line.						

Air, spark	N.			N.	C.	Band includes 2 lines.
Air, spark over water	Continuous spectrum and faint air-lines.					Band includes 2 lines.
Phosphoretted hydrogen	N.	Faint band.	Band includes			
Iron	VN.	N.	VN.	VN.		

Tested by coincidence, or close proximity of lines to those of the Aurora, we arrange the spectra in the following order:—(1) iron, (2) air-spark, (3) hydrogen, (4) air-tube, (5) phosphoretted hydrogen, (6) carbon and oxygen.

The air-tube spectrum might perhaps stand higher in the scale but for its broad bands, which make comparison doubtful. Lines of oxygen possibly escape detection in the Aurora from the faint character of its spectrum.

The phosphorus and iron spectra are especially interesting in connexion with Professor Nordenskiöld's "metallic and magnetic cosmic dust in the Polar regions" (see Phil. Mag. ser. 4, vol. xlviii. p. 546).

Additional Table of compared spectra.

As an addendum to the foregoing, on Plate IX. fig. 1 will be found a Table I have prepared, in which a type Aurora and also Vogel's and Barker's Auroræ are compared with eight other spectra, viz.:—

S.	Solar spectrum.
N.	Nitrogen (air): Watts.
O.	Oxygen (air): Watts.
C.H.	Carburetted-hydrogen vacuum-tube: Watts.
C.I.	Carburetted-hydrogen flame: Watts.
C.C.	Blue base of candle-flame: Capron.
O.P.	Oxygen vacuum-tube: Procter.
I.	Iron: Watts.

The divisions and vertical lines will guide the eye in making comparison of the spectra.

CHAPTER XII.

SOME NOTES ON PROFESSOR ÅNGSTRÖM'S THE-ORY OF THE AURORA-SPECTRUM.

[The substance of these appeared in the 'Philosophical Magazine' for April 1875, in conjunction with the "Comparison of the Tube and other Spectra" (Chapter XI.), but they are now, for the sake of convenience, made a separate article.]

Professor Ångström's propositions.

In a contribution by the late Professor Ångström to a solution of the problem of the Aurora-spectrum (an abstract of which appeared in 'Nature' of July 16, 1874), the Professor is stated, amongst other things, to have laid down certain propositions in substance as follows:—

That the Aurora has two spectra.

1st. That the Aurora has two different spectra—the one comprising the one bright line in the yellow-green only, and the other the remaining fainter lines.

That bright line does not coincide with HC_2.

2ndly. That the bright line falls within a group of hydrocarbon lines, but does not actually coincide with any prominent line of such group, and that Dr. Vogel's finding this line to coincide with a not well-marked band in the air-spectrum must be regarded as a case of accidental coincidence.

That moisture is nil in Aurora region.

3rdly. That moisture in the region of the Aurora must be regarded as *nil*, and that oxygen and hydrogen must alone there act as conductors of electricity.

Ångström's flask-experiment described.

Professor Ångström then details the examination of an exhausted dry air-flask filled with a discharge analogous to the glow of the negative pole of a vacuum air-tube.

Flask-spectrum compared with Aurora-spectrum.

The experiment is described as follows:—"Into a flask, the bottom of which is covered with a layer of phosphoric anhydride, the platinum wires are introduced, and the air is pumped out to a tension of only a few millimetres. If the inductive current of a Ruhmkorff coil be sent through the flask, the whole flask will be filled, as it were, with a violet light, which otherwise only proceeds from the negative pole, and from both electrodes a spectrum is obtained composed chiefly of shaded violet bands." The comparison of the spectrum of this violet glow with that of the Aurora gives, according to Ångström, the following results:—

Aurora-lines, wave-lengths	4286	4703	5226
Violet light, wave-lengths	4272	4707	5227

Two weak light bands, found by Dr. Vogel at 4663 and 4629, are also compared with other lines in the violet light 4654 and 4601; and the Professor then concludes that it may be in general assumed that the feeble bands of the Aurora-spectrum belong to the spectrum of the negative pole, possibly changed more or less by additions from the banded or the line air-spectrum.

Bright line is due to fluorescence or phosphorescence.

4thly. That the only probable explanation of the bright line is, that it owes its origin to fluorescence or phosphorescence. The Professor remarks on this point that "an electric discharge may easily be imagined which, though in itself of feeble light, may be rich in ultra-violet light, and therefore in a condition to cause a sufficiently strong fluorescence." He notes also that oxygen and some of its compounds are fluorescent.

No need of Dr. Vogel's theory of variability.

5thly. That there is no need, in order to account for the spectrum of the Aurora, to have recourse to the "very great variability of gas-spectra according to the varying circumstances of pressure and temperature" (Dr. Vogel's theory). Professor Ångström does not admit such variability, and does not admit that the way a gas may be brought to glow or burn can alter the nature of the spectrum.

Professor Ångström's conclusions tested.

In order to test some of the Professor's conclusions in an experimental way, I examined some tube and other spectra not only for line-positions, but also for general resemblance to an Aurora-spectrum.

These experiments are detailed in the last Chapter, and the results are comprised in Plates XIV., XV., and XVI., in which the spectra obtained are represented in black for white.

Result of examination of the Professor's propositions.

The result of the examination of Professor Ångström's principal propositions seems to be this:—

1st. Two Auroral spectra. I agree in this, but question whether the fainter lines may not possibly comprise more than one spectrum.

2nd. I agree also that the bright yellow-green line falls, as Professor Ångström describes, just behind the second line in the hydrocarbon yellow group (see Plate V. fig. 7). And I find, in common with the Professor, no well-marked or prominent line in the air-spectrum with which it accords.

3rd. This may be conveniently divided into two parts, viz.:—

A. The proposition that "moisture in the region of the Aurora must be regarded as *nil.*"

Moisture probably not nil in the Aurora region. Reasons for this given. Aurora in vapour or mist. Frequently near to earth's surface.

Here I see reason to differ, since (to quote a letter of Mr. Procter's) "the vapour-density of OH_2 is only 9 against 14 for N and 16 for O;" and again, "electrical or heat-repulsion may carry water-dust up to enormous heights." There are, too, I think, circumstances connected with the Aurora itself which make the assumption of moisture being *nil* in the Auroral regions untenable. The first of these is the fact that the white arc, streamers, and floating patches of light, found in some Auroræ, have frequently the peculiarly dense and solid look of vapour-clouds—a circumstance with which I have been frequently struck. Mr. Procter and others have also remarked that the Aurora is generally formed in a sort of "mist or imperfect vapour." The second, that Auroræ, or portions of them, are frequently near to the earth's surface. Instances of this are given in the section on the Height of the Aurora, notably the experiences of Sir W. Grove and Mr. W. Ladd.

Coincidence of Auroral lines with telluric solar lines.

On this point, too, note the peculiarities of the red line, which (and, as I find, the green line also) are coincident with, or very close to, telluric bands or groups of lines in the solar spectrum usually attributed to moisture. (See Plate XIII. fig. 2.)

Continuous spectrum.

I think we may also claim the continuous spectrum in the Aurora in further proof of water-vapour (see Plate XV. spectrum 7). The continuous spectrum of the Aurora is also, to my observation, more local and dense in the spectroscope than the glow generally seen between the lines or bands in gas-spectra.

Violet-pole spectrum discussed. Most spectra have a general as well as special character.

B. The question of the violet-pole spectrum. Here I make the remark that in comparing other spectra with that of the Aurora, it is, I think, too much the practice to trust to the coincidence (more or less perfect) of one or perhaps two lines out of many; whereas we know by experience that most spectra have so well-marked a general as well as special character that, when once seen, they are recognized afterwards with the greatest ease and without measurements. An experience and proof of this is found in a set of "Photographed Spectra" which the Autotype Company have reproduced for me.

Coincidence of one or two lines not sufficient to establish identity.

Of course no two given spectra can be considered identical unless their principal lines coincide; but, on the other hand, the coincidence of one or two lines out of many, without other features, cannot be satisfactorily or conclusively held to establish identity.

Ångström's compared spectra.

In Professor Herschel's letter (Phil. Mag. ser. 4, vol. xlix. p. 71), Professor Ångström's representation of the "spectrum of the glow discharge round the negative pole of air-vacuum tubes" is given, in comparison with the Aurora-lines and those of olefiant gas. This illustration is here introduced.

Ångström's representation of the Spectrum of the glow discharge round the negative pole of Air-vacuum tubes, and its comparison with the Spectrum of the Aurora.

Spectra of (1) olefiant gas, (2) Aurora, (3) negative pole in air.

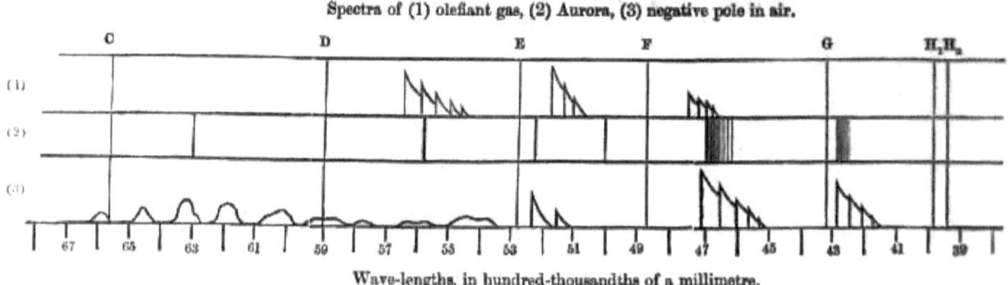

Wave-lengths, in hundred-thousandths of a millimetre.

It is unfortunate that in this illustration and in Professor Herschel's paper the wave-lengths of the Aurora-lines are not given in figures, but must be roughly calculated from the scale. Professor Herschel speaks of Ångström's drawing as representing a *normal* spectrum, and as derived from authentic sources, such as Vogel, Barker, and others; but beyond this we are not certain as to its origin.

In illustration of the difficulty of constructing any thing like a general typical Aurora-spectrum I append a Table of eight Auroral spectra taken at hazard:—

Table of compared Aurora.

Auroral lines and bands.

Observers.	Red.	Yellow.		Green.	Blue.	Indigo.		Violet.	
Vogel, April 9, 1871	6297	5569	5390	5233	5189	5004	—	4694 to 4629	—
Barker, Nov. 9, 1871	6230	5620	—	—	5170	5020	4820	—	—

						5050 to 4990	4930 to 4850	4740 to 4670	
Barker, Oct. 14, 1873	6300	5550	—	5330 to 5200	—	5050 to 4990	4930 to 4850	4740 to 4670	4310
A. Clarke, junr., Oct. 24, 1870		5690	—	5320	—	—	4850	—	4350
Backhouse, 1873	6060	5660	—	—	5165	5015	—	4625	4305
Backhouse, Feb. 4, 1874	*	5570	—	—	5180	4980	4830	4640	4320
H. R. Procter, 1870	*	*	—	*	—	—	—	*	*
Lord Lindsay, 1870	—	*	*	*	*	—	—	—	*

* Mr. Procter's and Lord Lindsay's lines had no wave-lengths.

Ångström's drawing discussed.

On examining Ångström's diagram it certainly seems to me that, upon the showing of the drawing itself, the coincidences are not very exact. All three of the violet-pole bands appear to be less refrangible than the Aurora-lines with which they are compared-the middle one (at 47) considerably so, the one near E (at about 52·30) appreciably so, and the third (at 43) slightly so.

Diagram of Vogel's Aurora and violet-pole spectrum.

As it seemed desirable to adopt a specific Aurora-spectrum for comparison, and to show such comparison on a somewhat larger scale than Ångström's drawing, I prepared the diagram shown on Plate XI. fig. 1. The upper spectrum is Vogel's, already described and figured on Plate XIII. The lower spectrum is that of "Air, violet pole," Plate XV. spectrum 2, graphically shown. I can only find one absolute coincidence in the two compared spectra in this diagram.

It should, too, I think, be borne in mind that there is a great difference in the character of the compared spectra, whether as shown in Ångström's drawing or mine—the bands of the violet-pole spectrum mostly degrading towards the violet, while the lines or bands of the Aurora in no way possess that character[14].

Dr. Vogel's violet-pole and Aurora-lines.

To assist in the foregoing violet-pole comparison I add the following Table derived from Dr. Vogel's memoir:—

---

[14] Ångström's drawing, in giving this character to the two Aurora-bands which are said to correspond with violet-pole bands about 47 and 43, is incorrect, and calculated to mislead by giving the Aurora-bands a feature corresponding to the violet-pole bands which they do not possess. I am not aware of any Aurora-line or band which is described as distinguished by degrading towards the violet.

Violet-pole lines.		Aurora-lines.	
W.L.		**W.L.**	
6100,	broad, moderately bright stripe	6297,	very bright stripe.
5945,			
5459,	broad, moderately bright stripe	5569,	brightest line of spectrum.
5289,		5390,	extremely faint line.
5224,	very bright line	5233,	moderately bright.
5147,	faint line	5189,	moderately bright.
5004,	bright line	5004,	very bright line.
4912,	fainter than last.		
4808,	very faint line.	6694,	band less brilliant in the middle.
4704,	very intense line.	4663,	
4646,	very faint line.	4629,	
4569,	moderately bright.		
4486,	moderately bright.		
4417,	quite faint line.		
4346,	moderately bright line.		
4275,	very bright line.		

On examination of these figures it will be seen that 5224 and 5233 are fairly close, and that 5004 is coincident. Beyond these there is little to identify the spectra.

Conclusions arrived at adverse to the violet-pole theory.

As the general result of my observations and a comparison of the foregoing spectra and tables, I see no reason for giving to the violet-pole glow any special or distinguished place in a comparison with the Aurora, and certainly not for assigning to it the nearly absolute monopoly of the spectrum. It is true that the line γ in the violet-pole glow (Plate XV. spectrum 2), which, by the way, degrades towards the red, is in close coincidence with one of the Aurora-lines; but how are the brighter bands α and β accounted for? These, as I have before pointed out, alone survive when the tube is placed at a distance from the slit. It is true they are thus reduced to shaded-off lines in lieu of bands; but the difficulty still remains, that they are conspicuous for their absence in the Aurora-spectrum. On the whole, I cannot but conclude that Professor Ångström's theory fails. At all events, if the violet-pole glow-spectrum is to represent the Aurora-spectrum, it must be under conditions different from those by which it obtains in dry-air vacuum-tubes or flasks at ordinary temperature.

Phosphorescence or fluorescence of the yellow-green line.

4th. I feel more in accord with Professor Ångström's memoir upon the subject of the phosphorescence or fluorescence of the bright yellow-green Aurora-line.

External features of Auroræ confirmatory of this.

I do not notice that the Professor touches upon the external features of the Aurora in respect of this question.

October 20, 1870.—I noted the grand Auroral display of that evening, including "streamers of opaque-white *phosphorescent* cloud very different from the more common transparent Auroral diverging streams of light."

February 4, 1872.—A fine display. The first signs were (in dull daylight) "a lurid tinge upon the clouds, which suggested the reflection of a distant fire, while, scattered among these, torn and broken masses of white vapour, *having a phosphorescent appearance*, reminded me of a similar appearance in October 1870." (Other instances of this effect will be found in the section Aurora and Phosphorescence.) Day Auroræ, too, we might suppose could hardly be seen without the presence of some phosphorescent glow.

Other confirmatory circumstances. Conclusion in favour of the theory.

Having regard to the near proximity of the phosphoretted-hydrogen band to the bright Aurora-line, to the circumstance of this band brightening by reduction of temperature (a phenomenon probably connected with ozone), to the peculiar brightening of one line in the green in the "Aurora" and "phosphorescent" tubes (the phosphorescent tubes probably containing O), and to the observed circumstance that the electric discharge has a phosphorescent or fluorescent after-glow (isolated, I believe, by Faraday), I feel there is strong evidence in favour of such an origin to the principal Aurora-line, if not to the red line as well.

Invariability of gas-spectra questioned.

5th. Professor Ångström opens a wide door to discussion in his proposition of the invariability of gas-spectra, and I do not now attempt to follow in detail this interesting part of the present subject. Suffice it to say, that if the Professor lays down this proposition in its strictest sense (I can hardly suppose he so meant it), there is, so far as I am aware, no one spectrum that can at all claim comparison with the Aurora-spectrum. Giving greater latitude to the Professor's words, I reply, upon competent authority, that lines vary in number and brilliancy with temperature, and in breadth with pressure. Kirchhoff, too, in speaking of vapour-films as increasing the intensity of lines, states "it may happen that the spectrum appears to be totally changed when the mass of vapour is altered." We may, too, now add magnetism as capable of effecting a change in certain spectra, not only as to brilliancy, but even as to position of lines. (Chautard's Researches, 'Philosophical Magazine,' 4th series, vol. 1. p. 77, and experiments detailed in Chap. III. of this work.)

CHAPTER XIII.

THE OXYGEN-SPECTRUM IN RELATION TO THE AURORA (PROCTER AND SCHUSTER).

Procter's oxygen-spectrum.

In a communication to 'Nature,' Mr. H. R. Procter has pointed out an apparent coincidence in position of several of the Auroral lines with those of a spectrum occasionally obtained from air at low pressure with a feeble discharge. It is, he says, sometimes exhibited in lumière (phosphorescent?) tubes, and he believed it, in part at least, to be the spectrum described by Wüllner (Philosophical Magazine, June 1869) as a new spectrum of oxygen.

How obtained.

He had obtained it very vividly in pure electrolyzed oxygen with a feeble discharge, but some perplexing observations made him doubtful of its origin.

Plate XI. fig. 4 gives a representation of this spectrum as shown by Mr. Procter, except that my drawing is in black for white.

Compared spectra described.

The upper spectrum is that above mentioned, the centre one that of the Aurora, the lower one the lines of Na and H for comparison. The Auroral yellow-green line, in January 1870, was found by Mr. Procter coincident with a bright line or band in the tube (with a spectroscope of a 60° bisulphide prism, and magnifying-power about six). The third and fifth lines in the Aurora seemed also to correspond with tube-lines. As to these Mr. Procter says they were not bright enough to be compared with the same accuracy as the yellow-green line, but that the positions could not be far wrong.

Mr. Procter's subsequent views. Yellow-green line traced to some form of hydrocarbon.

Mr. Procter subsequently ('Edinburgh Encyclopædia,' art. "Aurora") considered he traced the yellow-green tube-line to some form of hydrocarbon. On examination with instruments of greater dispersion, it was found that, though more refrangible than the first band of citron acetylene (candle-flame), it was less so than the Aurora-line. The tube-band, too, was shaded towards the violet, which was not the case with the Aurora-line.

The question as between hydrocarbon and oxygen I did not then consider as disposed of. With the lumière tubes the question might be open, but I did not see how it could be in the case of the electrolyzed oxygen-spectrum.

From a comparison of the tube-spectra, I have shown that although the spectra of the carbon and oxygen tubes are proved to be, photographically, as a whole, distinct, they have, as to position of some of the principal lines in the central part of the spectrum, a very close resemblance.

Probability that O may play a part in the Aurora-spectrum.

That oxygen may in some form play a part in the Aurora seems highly probable; how far it is spectroscopically detected seems a different question.

Difference between air-spark and tube-spectra.

Ångström and Herschel suggest its presence in the Aurora in connexion with phosphorescence or fluorescence. With a spark-discharge in air at ordinary pressure, a mixed spectrum of bright lines of N and O is found; while in the case of Geissler vacuum-tubes (representing a glow-discharge in a much more rarefied atmosphere) the N lines appear mainly to usurp the spectrum.

H_2O tube referred to.

It must, however, be borne in mind that a Geissler tube, as to temperature at least, in no way represents the conditions of the Aurora; and when we remember the association of oxygen and ozone, and the way in which the latter is affected by heat, it may well be that temperature plays an important part in the matter. In proof of this conduct of oxygen, it may be cited that, in the case of a H_2O tube, the H lines come out sharp and brilliant in the spectrum, while what is seen of the O lines is comparatively weak, misty, and ill-defined. Vogel, it will be remembered, makes 5189 of the Aurora coincident with an O line.

Residual phosphorescence in Geissler tubes. Garland tube.

Professor Herschel has pointed out, and I have confirmed, that the residual phosphorescence in Geissler tubes, after the spark has passed, is probably associated with oxygen. He also alludes to the fact that when one of the globes of a "Garland" tube was heated, it did not shine after the spark had passed, apparently because of the destruction of the ozone by heat.

[Some experiments with a tube of this description will be found detailed in Part III. Oxygen was not, I think, the gas it was filled with.]

Dr. Schuster's tubes described.

Subsequently to my examination and comparison of the O and CO_2 spectra before detailed, Dr. Arthur Schuster was good enough to send me three vacuum-tubes of his own preparation, showing an oxygen-spectrum.

One, with large disk-shaped brass electrodes, was unfortunately broken in transit. Dr. Schuster informed me it showed the carbonic-oxide spectrum as well as that of oxygen. The other two tubes had aluminium electrodes. They were similar in shape to ordinary Geissler tubes, but had attached to each a supplemental bulb containing dry oxide of manganese. Illuminated by the larger coil, one of these tubes (which had a slight crack in the manganese bulb) lighted up faintly; the other was fairly bright, and the glow had a somewhat reddish tint.

Plate XVIII. fig. 15 represents as the upper spectrum Vogel's Aurora, with W.L. numbers, as the middle spectrum the capillary part of Dr. Schuster's O tube, and as the lower spectrum the negative (violet) pole of the same tube.

Spectra described.

The tube-spectra were mapped out with the aid of the diaphragm micrometer before described.

Capillary.

The capillary spectrum was mainly distinguished by four bright sharp lines — one in the red, between the red Aurora-line and D, two in the green, but considerably more refrangible than the yellow-green Aurora-line, while the fourth was found to be hydrogen F. The other lines in the spectrum were considerably fainter, and misty and band-like. The red line, though not brilliant, was fairly bright and sharp.

The place of the less refrangible of the two bright bands in the violet-pole spectrum was occupied in the capillary spectrum by a faint glow only.

Violet-pole.

The violet-pole spectrum was recognized by two very bright broad bands of light in the green, each including within its limits one of the Aurora-lines. The bright red line in the capillary had a faint representative in the violet-pole spectrum, as also had the two bright lines in the green. Other fainter lines appeared in the blue, and three fairly bright ones towards the violet.

Dr. Schuster's remarks on the spectra.

Dr. Schuster remarks that one of these O bright bands is closely coincident with a band in the CO spectrum, but that the CO band is bright towards one edge and fades off gradually thence, while the O band is of pretty uniform strength throughout. Dr. Schuster finds the wave-lengths of the violet-pole O bands to be as follows:—

$$\left.\begin{array}{l} 5205\cdot0 \\ 5292\cdot5 \end{array}\right\} \text{Brightest part } 5255.$$

$$\left.\begin{array}{l} 5552\cdot8 \\ 5629\cdot6 \end{array}\right\} \text{Brightest part } 5586.$$

His tubes free from impurity.

He also gives as weak bands 5840-5900 and 5969-6010. Dr. Schuster comes to the conclusion that the green line of the Aurora is not due to oxygen, as, under considerable dispersion and with good definition, the oxygen-bands can be broken up into a series of lines, when the brightest part is found to lie at 5586, which is too much towards the red to compare with the Aurora-line. He notices that the more refrangible of the O bands corresponds with a line sometimes seen in the Aurora (Vogel's 5233). The same remark will, however, apply to this last as to the other coincidence, viz., that a broad band can hardly represent a line—at least, the line can only be said to coincide in a loose and indefinite way. It is evident that Dr. Schuster's tubes were free from what must now be considered an impurity in those examined by me and by Dr. Vogel, and that Mr. Procter's suspicions of carbon impurities in these, and the ordinary oxygen-tubes, are thereby quite confirmed.

Experiments with an open Geissler tube.

In some experiments which we made (after receiving Dr. Schuster's tubes) with an open Geissler tube, so arranged as to connect with an air-pump and gas-receiver, and thus from time to time to wash out the tube and vary its contents, we found the same impure spectrum as in the case of the sealed O tubes; and it seems to require a very large amount of precaution to avoid these impurities.

Spectra of Dr. Schuster's O tube examined.

Dr. Schuster was kind enough to examine the spectra I mapped out, and which are shown in Plate XVIII. fig. 15, with the following results:—The lines Oα, Oβ, Oγ are those he has referred to under that designation in his communications to 'Nature,' and undoubtedly belong to oxygen. The bands A, B, and C are the bands characteristic of the negative pole. He finds A divided into two parts by a dark space. The spectrum of the negative pole, under good exhaustion, stretches into the capillary part; hence B appears in the capillary as a faint band. A similar thing happens with nitrogen. I., II., III., and possibly 8 and 9, he thinks, are due to the spark-spectrum of oxygen, obtained when the jar and a break are interposed, the brighter lines of the line-spectrum being always present at the negative pole. These last-mentioned lines I have already referred to, as having been found by me in a tube showing phosphorescence after the spark has passed. (Compare Plate XVIII. fig. 15, O violet pole, with Plate XV. spectrum 5.) Nos. 1 and 2, he thinks, are due to some foreign matter, as they are not in all his tubes.

Dr. Schuster often finds that a spectrum due to the aluminium electrodes is seen in tubes under great exhaustion; and this he considers is the spectrum of aluminium oxide. A drawing of this spectrum is found in Watts's 'Index of Spectra,' plate iii., "Aluminium first Spectrum." To this, he thinks, are also due the bands, or sets of lines in my aluminium-arc spectrum ('Photographed Spectra,' plate ii.), and he believes lines 3, 4, 5, 6, and 7 in the mapped-out spectra are due to it. It would thus appear that the lines due to O are few in number, and do not well compare with the Aurora-spectrum.

PART III.
MAGNETO-ELECTRIC EXPERIMENTS IN CONNEXION WITH THE AURORA.

INTRODUCTION.

Object of experiments. Description of apparatus employed. Electro-magnet. Battery.

The set of experiments detailed in Chapters XIV. to XIX. was mainly conducted for the purpose of testing, in connexion with the Aurora, the action of a magnet upon the electric glow *in vacuo* and on the spark at ordinary pressure. It also includes some observations on the glow from the violet pole with and without the magnet, and on the glow obtained from one wire only. The apparatus employed was a Ladd's electro-magnet, with poles 10¼ inches high by 2 inches across, each pole being surrounded by a movable helix, composed of two sets of stout copper wire wound together, so that they could be used either in one length or as independent coils excited at the same time. The latter form of arrangement was employed by us. In most of the experiments conical armatures were employed for the purpose of bringing the action of the poles to bear upon the subjects examined. A contact-maker was added to the magnet, so that it could be put rapidly in or out of action without disturbing the wires. The battery used to excite the magnet was of the form known as that of Dr. Huggins, and consisted of four vulcanite cells in a frame, each holding seven pints of bichromate solution, and containing two carbon and one zinc plate, each 13½ by 6 inches.

Small coils. Larger coil. Magnetic curves obtained.

A winch and pulley enabled the whole set of plates to be lowered into the liquid and withdrawn at pleasure, and the large quantity of solution gave the battery a considerable amount of constancy. We found it could be used for two evenings' work, of four hours each, without any material dropping in power. For obtaining the glow in the Geissler and other small tubes, a Ruhmkorff coil, giving a ½-inch spark, excited by one plate of a ½-gallon bichromate (bottle form), was used. For the glow in the larger tubes and the spark in air a larger coil, giving a two-to three-inch spark, and worked by two ½-gallon double-plate bichromates, was employed. Notes were taken of the experiments, and drawings of the effects at the time; and these are reproduced almost literally in the text and Plates comprised in this Part. To ascertain the direction and extent of the magnetic curves, we covered large sheets of cardboard, placed over the poles, with iron filings; excited the magnet so as to obtain the curves, and then obtained permanent prints from the filings by spraying the cardboard with tannin solution. The magnetic effects were thus found to extend to a radius of at least ten inches (see diagram, Plate XVII. fig. 1, showing magnet-poles and curves on a ¼ scale).

Chautard's investigations kept in view. Evidence obtained of change in colour, form, &c.

of Aurora. Ångström's flask-experiment tried.

In the vacuum-tube experiments we held Mons. J. Chautard's investigations (on the action of magnets on rarefied gases in capillary tubes rendered luminous by the induced current, Phil. Mag. 4th series, vol. 1. p. 77) in view. We obtained in our experiments plenty of evidence of a change of colour and form in the discharge under the magnetic influence; and both simple and compound spectra were found to be much varied by the exaltation or suppression of some parts of the spectrum, so that apparently new lines sprang up; but we failed to trace actual change of position or wave-length in any given line, though we carefully looked for it. A portion of our researches was directed to the subject of Ångström's experiment of filling a dry flask with a violet glow, analogous to that from the negative pole. We entirely failed in obtaining the same result while two wires and an uninterrupted circuit were employed. When, however, we attached a negative wire only (the other wire being left free) to an exhausted globular receiver, we obtained an effect very similar to that referred to in Prof. Ångström's memoir.

General results of experiments. Bulb effects noticed as a mode of analysis of gases.

The general result of the experiments was to prove, assuming the Aurora to be an electric discharge, the great influence the magnetic forces may exercise on the colours, form, motions, and probably the spectrum also of that phenomenon. It is easy to conceive that the variation in number, and intensity of the lines which has been remarked in Auroral spectra may have its origin in such a cause. The influence of the magnet on the capillary stream was mainly in colour and intensity; but in the bulbs the effects were still more marked and striking, and, in a greater or less degree, different in the case of each gas which we examined. A careful and extended study of these effects, conjointly with the changes in the spectrum, might possibly form a new and valuable mode of analysis of compound gases. This is well illustrated in the case of the iodine and sulphur tubes which we examined.

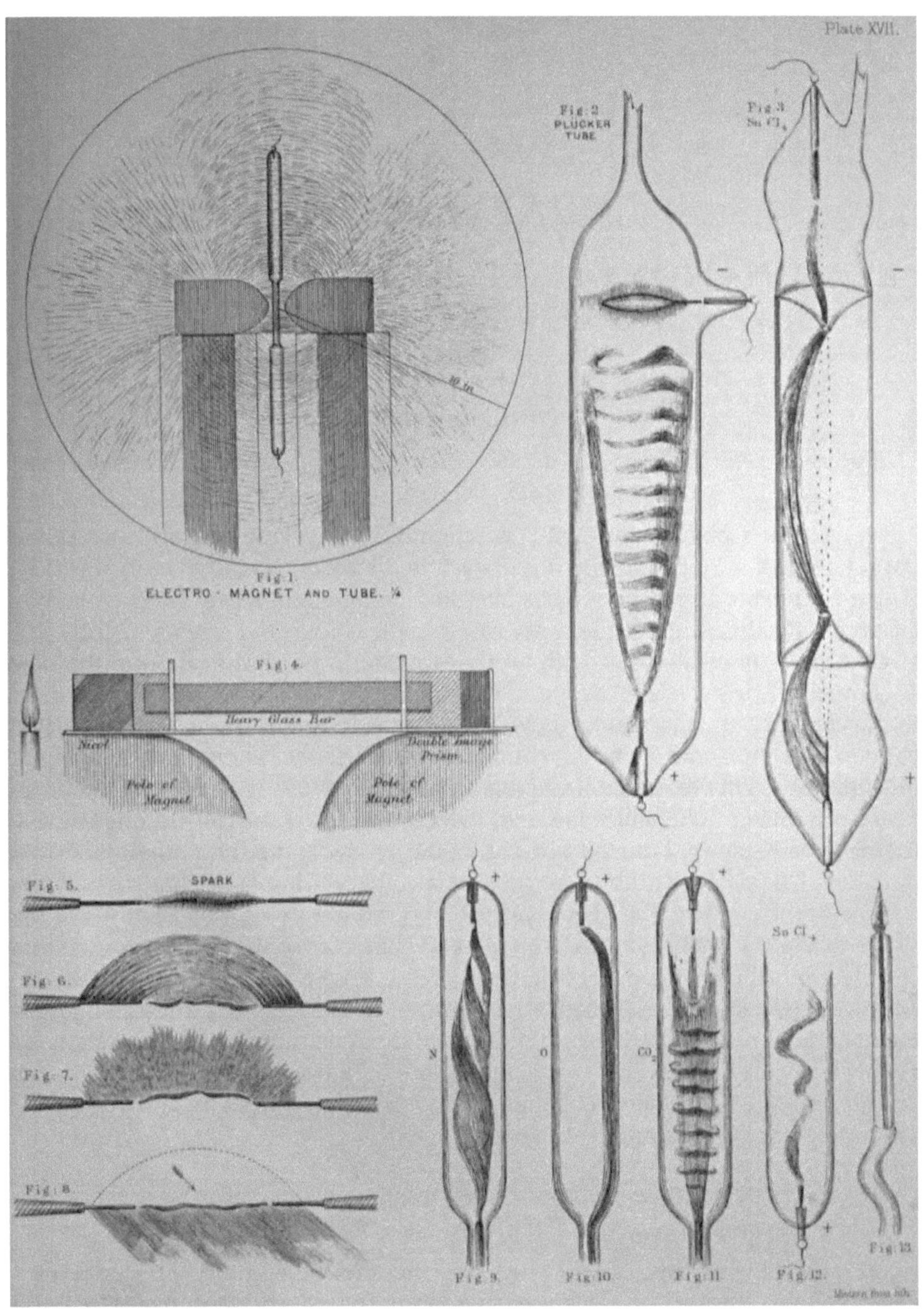

Plate XVII.

EFFECT OF MAGNET ON TUBES AND SPARK

CHAPTER XIV.

EXAMINATION OF GEISSLER TUBES UNDER ACTION OF THE MAGNET.

Nitrogen-tubes.

Nitrogen-tube No. 1. Discharge described. Spectrum described. Capillary stream. Positive bulb. Violet-pole glow.

(1) A small Geissler tube (No. 1) was lighted up by the small coil. The capillary part showed a very bright, slightly rosy-tinted stream. Negative bulb was filled with rosy-purple light, the violet-pole glow being confined to the extent of the electrode. Positive bulb of the same rosy-purple colour, but stream slightly contracted in volume. Glow throughout quiescent, and no stratification in the tube. A compound-prism spectroscope, taking in the whole of the spectrum, showed in the capillary stream, from yellow to red, a fairly bright wedge, having a dark band in the centre, and six bright columns, with dark lines at intervals, shading off on either side. On the more refrangible side of the yellow, the spectrum was composed of a set of bright bands and lines in the green, blue, and purple, one line only (in the green) standing out very bright. In the yellow and red no bright line stood out alone. The positive bulb gave a fainter spectrum of the same character, mainly confined to the centre, the violet, yellow, and red not being well seen. When the violet-pole glow was examined, the general character of the spectrum was quite changed: a brilliant broad band in the violet, a bright narrower one in the blue, and two bright lines in the green, with intermediate fainter lines throughout, were the main features. The yellow and red part of the spectrum was also changed. The yellow was fairly and evenly distinct up to the dark band; then came a somewhat brighter orange band, and after that the red, but rather obscure and cut off. No absolutely bright line could be traced in the red.

Nitrogen-tube No. 2. Glow described. Difference of spectra of capillary stream and violet-pole glow. Junction of the violet-pole glow and capillary stream.

(2) To compare the capillary stream and the violet-glow, a second nitrogen-tube (No. 2) was used. This tube was larger in bulk and bore than No. 1. The glow in the bulbs was considerably fainter and more salmon-coloured; and there was much stratification in both, extending to the capillary bore. (This stratification was considered due to H, as the three principal lines of that gas came out very brightly in the spectrum.) The difference of the spectra of the capillary stream and

of the violet-pole glow was extremely well marked—the former consisting of a set of bright lines and bands of fairly uniform intensity, while the latter was split up into a few bright bands with fainter lines between. The yellow and red of the violet-glow were very weak as compared with the same region of the capillary spectrum. No bright line appeared in the red. The tube being properly adjusted for the purpose, the junction of the violet-pole glow and the capillary red-glow was easily observed. The bright bands of the violet-pole were seen to run into the capillary line-spectrum, and then, gradually getting finer and more pointed, to fade out.

Tube No. 1 between the poles of the magnet. Change of colour in capillary stream. "Tailing-over" of capillary stream.

(3) The capillary part of tube No. 1 was arranged between the poles of Ladd's electro-magnet, the conical ends of the armatures almost touching the tube (Plate XVII. fig. 1). With the magnet not excited, the capillary stream was bright and of a slightly rosy-yellow tinge. It varied a little in apparent diameter with the current. As soon as the magnet was excited the capillary stream, as also (in a less degree) that in the bulbs, were seen to contract, and to change from a *rosy* tint to a distinctly *blue-violet*. The polished armatures, acting as reflectors, showed this change of tint in a most marked manner each time the magnet was excited. At the same time the capillary stream was seen to run into the negative bulb, as if overflowing, and with an effect resembling the "tailing-over" of a gas-flame. This effect took place each time the magnet was excited, and was not found at the positive-bulb end.

Spectrum examined.

Occasionally, when the magnet was excited, flashes of light were discharged in the negative bulb from the capillary towards the violet-pole. The spectrum was then carefully examined. No change was seen in the actual position of any of the lines or bands when the tube was influenced by the magnet, but those towards the violet end of the spectrum were conspicuously brightened.

Negative bulb between poles of the magnet. Positive bulb within action of the magnet.

(4) The extremity of the negative bulb was now placed between the poles of the magnet. A bright violet-coloured arc, following the magnetic curve, was at once formed, as in the case of the large Plücker tubes; and at the same time a straight stream of not very bright light ran along the bulb. The positive bulb was next placed within the action of the magnet; and immediately a brilliant spiral of flickering light appeared in the bulb, lighting it up, and reminding one in shape of the spiral which water forms on being poured from a lipped jug (see Plate XVII. fig. 9).

Spiral formed.

This was repeated each time the magnet was excited. The spiral, though flickering in character, was permanent in form, and inclined to the side of the tube which was in contact with the N pole of the magnet.

Oxygen-tubes.

O tube No. 1; spectrum described.

A tube (No. 1) was lighted up and examined with the spectroscope, and found to give the spectrum shown on Plate XIV. spectrum 3, but with a strong set of H lines in addition.

O tube No. 2; spectrum described.

A second tube (No. 2) was then lighted up. The spectrum was a bright one, similar to the foregoing, the principal H lines being present, but not strong.

Tube-glow described. Effect upon glow when magnet excited. Bulbs between poles of the magnet. Effect of magnet on spectrum.

The red region was indistinct, and showed no prominently bright line. The bulbs were mainly of a slightly blue-grey tint, with a steady glow. Capillary stream quite pale white, with a very slight tinge of red. Violet-glow small and confined to the electrode. Upon the magnet being excited, the capillary stream became intensely brighter, and the glow in both bulbs contracted into a single bright stream, which curved towards the sides of the bulbs at right angles to the magnetic poles, and changed from side to side with the current. This effect was very marked, and was more apparent in the positive than the negative pole. A faint stratification was seen in both bulbs. Upon either bulb being placed between the armatures, the glow left the electrode point and condensed into one bright stream, running along the side of the tube and curving at each end (Plate XVII. fig. 10). No trace whatever of tendency to form a spiral was seen. The spectrum with the magnet on was very conspicuously brightened up throughout. A set of fluted bands with a bright line among them appeared in the red, and several lines or bands appeared in the violet which could not be seen before. The bright red line, upon measurement, proved to be the hydrogen-line C. It thus seemed brighter in proportion than the F line, although, with the magnet off, the latter was well seen, while the C was not. No actual change in position of the spectrum-lines could be detected.

[It is to be noticed that the O tubes employed were those used by me in former experiments, and had the bright lines now attributed to hydrocarbon impurity. Their bulb-effects differed, however, entirely from those of the CO_2 tube. (Compare figs. 10 and 11, Plate XVII.)]

Hydrogen-tubes.

H tube, No. 1; glow described.

A small H Geissler tube (No. 1) was selected, and lighted up by the small coil. The capillary was a bright white-pink stream, with a tendency to redden at times. The bulbs were both of a faint blue-grey tint, with coarse lenticular stratification. The violet-pole glow was pale and white as compared with that of N.

When the magnet was excited, whole character of tube changed. Unexcited spectrum described. Effect when magnet was excited.

When the magnet was excited, the whole character of the tube changed. The capillary stream diminished in brightness and in apparent volume, and changed to a deep amber-yellow. The bulbs lost some of their light, and their coarse stratification; being, in lieu, filled with a vertical condensed stream of moderate light, in which a fine stratification only was seen. The stream in the positive bulb had a tendency to the spiral form. The capillary, each time the magnet was excited, "tailed over" into the negative bulb, as in the case of N, looking as if it were squeezed out of the capillary bore. The unexcited spectrum was found to consist of the usual principal lines of H on a continuous glow, with the intermediate bands and finer lines, which are usually suspected to be due to impurity. The sodium-line was also seen. When the magnet was excited, the spectrum grew much fainter—the continuous glow in the red and blue, and the red and blue lines, nearly disappearing, and the line in the green alone shining out conspicuously. No change of place in the lines could be noticed.

No. 2 H tube; effects described.

A longer H tube (No. 2) was then tried, with similar effects, except that the diminution in brightness was not so conspicuous. When the negative bulb of the tube No. 1 was placed between the poles of the magnet, a stream of light was formed, and the stratification became finer. The same effect took place with the positive bulb, with a tendency to the spiral form.

Water-Gas (H$_2$O) tube.

Water-gas (H$_2$O) tube; effects produced described.

A faint purple glow was seen in each bulb, the tube not lighting-up brightly. The capillary showed a slightly rosy-tinted, grey stream of brighter light. With the magnet on, the glow in the bulbs was condensed into a single bright stream. The capillary brightened up, and assumed a yellow tint—this effect being principally confined to that portion which was between the conical ends of the armatures, and gradually diminishing as the distance increased from these. Without the magnet, the principal H lines showed brightly in the spectrum, the O lines being misty and indistinct. With the magnet on, the O lines and spectrum generally brightened up.

Ammonia-tube.

Ammonia-tube; lighting-up described. Spectrum described.

This tube was difficult to light up. Hardly any light was seen in the bulbs, except a very faint purple glow at the electrodes. In the capillary part a fairly bright stream of purple-white light appeared. The spectrum was a faintly shown one of N and H. The effect of the magnet was to reduce the brightness of the glow in the capillary, but with little marked action on the bulbs, except to condense the faint glow into a slightly bright stream running along the side of the tube.

On a subsequent examination the tube and spectrum both brightened up under the influence of the magnet. The N lines, which were faint without the magnet, shone out under its influence distinctly—the red and yellow parts of the spectrum specially showing this effect. The H lines also brightened up, but hardly so much in proportion as the N.

Carbonic-Acid tube.

Tube marked C A; lighting-up described.

A Geissler tube marked C A was examined. Capillary stream a brilliant bluish white; bulbs grey-blue, with a slight tint of green; slight stratification in positive bulb; stream diffuse, not quite filling the bulbs, and changing in volume as the coil-break was touched; glow round the violet-pole considerable, but markedly white in tint, rather than violet; stratification strong in capillary. With magnet excited, the capillary stream diminished in volume, but greatly increased in brightness. It "tailed over" into the negative bulb, and the stream through both bulbs curved towards the sides. A slight pattering noise was heard in the tube. In the positive bulb bright, imperfectly formed, saddle-shaped rings of light, with a tendency to spiral formation, were seen, somewhat similar to the effects in the Plücker tube after described (see Plate XVII. fig. 11).

Effects when magnet was excited.

The whole spectrum, under influence of the magnet, became much brightened up. Faint bands in the red came out bright, as also did some in the violet. The violet-glow was examined (without the magnet), and the light was found condensed into four prominent shaded bands, one red, one yellow-green, one green, and one blue, with fainter bands seen between.

Chlorine-tubes.

Chlorine-tube No. 1 lighted-up. Action of magnet upon the tube and spectrum.

A chlorine-tube (No. 1) was lighted-up with the small-coil. Capillary stream of a pale green tint. Bulbs with very little glow in them; spectrum pale, and not very distinct. Under action of the magnet this tube brightened up throughout, and the glow became more condensed, and ran to the sides of the tube. The spectrum also brightened, the faint lines becoming stronger, but the general character was preserved.

Chlorine-tube No. 2 lighted-up. Effect on glow when magnet was put on.

A second chlorine-tube (No. 2) was then tried. Both bulbs were completely filled with a dense white (very slightly rosy-tinted) opaque light, and capillary the same, but brighter. A very slight violet tinge was seen at the negative pole. When the magnet was put on, both bulbs were at once filled with flickering bright streams of light, running towards the side of the tube, according to the direction of the current.

The capillary stream at the same time changed from white to an intense bright green. The spectrum without the magnet consisted of sets of lines, with two well-marked absorption-spaces between, all seen somewhat faintly, as if through a mist.

Changes in spectrum when magnet was excited.

When the magnet was put on, the marked character of the absorption spaces was lost. The sets of lines in the yellow-green and green started up intensely bright, while those in the blue only slightly brightened.

The misty appearance was altogether lost, and the bright lines all shone up upon a perfectly dark background, with a strikingly metallic look; we could not, however, trace change of position or actually new lines. It seemed as if lines which had been faint in the yellow-green and green region suddenly increased in intensity, the other parts of the spectrum not being similarly influenced. They quite flashed up when sudden contact was made with the magnet commutator.

Iodine-tubes.

Iodine-tube No. 1.

This tube (No. 1) had been used for photographic purposes, and the bulbs were partly obscured by a white deposit.

Lighting-up described. Effect of touching one wire with the finger.

On lighting it up, both bulbs were filled with a violet-grey diffused light, with much coarse well-marked lenticular stratification. This stratification was mainly lost on changing the direction of the current, but made its reappearance when one conducting-wire was touched with a finger. This effect was still more marked when one finger of each hand was applied to the wire. The capillary stream was of a pale lemon-yellow. On putting on the magnet the light in the whole tube was nearly extinguished, a faint thin stream of condensed light running through the centre of the tube alone remaining.

Effect of the magnet.

On placing the bulbs between the magnet-poles, effects were produced similar to those in the case of the tube, after described (p. 144, and marked Si Fl$_6$), but in a less marked degree.

Tube again tested.

The iodine-tube was subsequently again tested, and it lighted-up better than on the last occasion, showing nearly the same effects in bulbs and capillary, the former having somewhat of a rosy tint and the latter an amber.

Magnet effects. The spectrum described. Change when magnet was excited.

On exciting the magnet, the capillary part of the tube changed from amber to a decided light green. The spectrum, without the magnet, gave one very bright line, and several less bright ones near, in the blue-green. The rest of the spectrum, with the exception of the absorption-spaces, was misty and continuous, with lines showing faintly through. The red and yellow portions of the spectrum were quite bright. When the magnet was excited, the spectrum entirely changed. The red and yellow portions of the spectrum, and the misty continuous light, all quite disappeared; while a set of sharp lines on the yellow-green and green flashed up bright and clear, and stood out alone upon a dark background, in which the absorption-spaces were lost. The effect was very strongly marked, and gave a totally different character to the appearance of the spectrum. The change seemed to arise from the suppression of one part of the spectrum, and the increase in intensity of the lines in the other part.

No change in line-position.

The principal lines could not be traced to change in actual position.

This tube differing somewhat from a second one we examined (No. 2) in tint of glow and spectrum, it suggested itself to us that there might be a partial mixture of N or H (or both) with the iodine vapour, giving rise to some of the brighter parts of the spectrum which were extinguished under the action of the magnet.

Comparison of iodine-tubes No. 1 and No. 2. Comparison of the spectra.

We therefore compared these two tubes, viz. the old one (No. 1) and the new one (No. 2), and also their spectra, by means of a comparison-prism on the slit of the spectroscope. To the eye, the tubes differed much in appearance. No. 1 had a distinct transparent rosy tint throughout, with considerable coarse flickering stratification; and this contrasted strongly with the dense whitish light of tube No. 2, which showed neither movement nor stratification. The spectra were also found different in general look. That of tube No. 1 was strongly tinged in the red and yellow, and showed a bright continuous spectrum, crossed by many sharp lines, with little trace of absorption-spaces. The spectrum of No. 2 was much whiter in tint, showed very little of the red and yellow, and the absorption-spaces were very dark. A few bright lines, mainly in the yellow-green and green, were faintly seen.

The two tubes examined in detail. No. 2.

The two tubes were then examined separately in detail. No. 2, excited by the magnet, showed curious effects. The glow was rendered weak and intermittent, and the rosy tint almost disappeared. The capillary changed to a decided green colour, and the positive electrode was surrounded by a yellow glow. The changes in the spectrum were no less decided. Without the magnet, the spectrum was found to be a bright continuous one of H (with a full set of principal and intermediate lines) and N—the N spectrum being rather faint and misty, with very slight, if any, traces of the iodine-spectrum. On the magnet being excited, the spectrum changed as if by magic; the H and N spectra disappeared (except hydrogen F, which still faintly remained), and the iodine lines, mostly in the yellow-green and green, shone out wonderfully sharp and bright on quite a dark ground. No. 1, upon examination, showed between the magnet-poles only the same changes as on last occasion. The spectrum seemed to be one of iodine, with the addition of slight traces of the H spectrum.

Effects discussed.

On excitation of the magnet, the misty continuous part of the spectrum nearly disappeared, and the bright lines shone up sharply upon the dark background as before. The effects in the case of both tubes were strongly marked. The impression as to tube No. 2 was that, without the magnet, the slight iodine-spectrum was overpowered and masked by the N and H spectra; while under the influence of the magnet the N and H spectra were almost altogether suppressed, the iodine-spectrum being at the same time intensified. The disappearance of the continuous spectrum under the action of the magnet in No. 1 (with the supposition it was mainly H) would be accounted for in the same way.

Bromine-tubes.

Bromine-tube No. 1. Lighting-up described. Effect of the magnet. Bromine-tube No. 2. Effect of magnet.

This tube (No. 1) had been previously worked for photographic purposes. Excited by the small coil, the whole tube was filled with a faint flickering light. The positive bulb contained a faint purple glow, with a yellow-green tinge at the electrode, a curious flickering stream of light flashing from the electrode to the side of the tube. The negative pole showed pretty much the same effect as the positive. The capillary stream expanded at the opening into the positive bulb, but ran in a condensed stream into the negative bulb. In colour it was of a rather bright lilac. Upon putting the magnet on, the light-glow in the tube was at once and permanently extinguished, the coil still working as if the current passed. The same effect happened repeatedly; but now and then the tube lighted-up for a second, showing spiral arrangement in the bulb. We tried another bromine-tube (No. 2): it lighted-up easily; both bulbs were filled with a purple stream of light; capillary stream bright grey. The glass of the tube was strongly fluorescent and of a yellow tinge. When the magnet was excited the stream of light was somewhat condensed in the bulbs, and flew to the side of the tube; while the capillary stream at the same time brightened. The spectrum without the magnet was fairly bright; it increased in brightness under the influence of the magnet, and additional lines appeared; but we considered them to be only faint existing ones brightened up. No change in the position of the principal lines was traced.

Silicic-Fluoride tubes.

Si Fl$_6$ tube. Lighting-up described. Effect of magnet.

(1) A tube marked Si Fl$_6$ had been worked for photographic purposes; it lighted-up easily. Both bulbs were filled with a brown-pink diffused light, inclined to condense into a stream in the positive bulb. The violet glow was very bright, and nearly filled the space round the electrode. The capillary stream was of a bright violet tint. The effect of the magnet was to decrease the intensity of the light throughout the whole tube.

In the positive bulb the stream broke up into a number of vibrating streamlets, with little bright threads of light intermixed, which flew towards the side of the tube at right angles to the magnetic poles. There was an inclination to spiral arrangement in the streamlets. This stream changed from side to side of the tube coincidently with change in the magnetic poles. At the negative pole the violet glow formed an arc in the direction of the magnetic curves, while a spiral of fainter (positive?) light was formed in the upper part of the bulb. A slight ringing sound was heard in the tube.

Comparison of Si F_4 and Si Fl_6 tubes.

(2) We compared two tubes (Si F_4 and the one marked Si Fl_6). The Si Fl_6 tube in general effect, and in its spectrum, when lighted-up, resembled Si F_4. We compared the one tube under the influence of the magnet with the other not so, by means of a comparison-prism on the slit. As the spectroscope and second tube were necessarily removed some distance from the magnet, the spectrum of the tube between the poles was not bright. We could not trace a change of position in any of the principal lines. The tube between the poles was brightened up when the magnet was in action[15].

Sulphuric-Acid (SO_3) tubes.

SO_3 tube No. 1; lighting-up described. Effect of the magnet. Changes in the spectrum.

(1) Excited by the small coil, both bulbs of this tube (No. 1) lighted-up brightly, with a misty light-blue tinted stream of opaque light, a yellow glow appearing at the negative pole. The capillary stream partook of the same blue tint, but was whiter and brighter. Under the magnet's influence, the glow in the bulbs flew to the side of the tube in flickering streams of light, the capillary at the same time changing to a distinctly green tint. The spectrum without the magnet consisted of four fairly bright bands of light in the yellow, green, blue, and violet, connected by a faint misty continuous spectrum (O or possibly the hydrocarbon spectrum found in O tubes by way of impurity).

When the magnet was excited, this spectrum entirely disappeared; and a set of bright metallic-looking lines upon a dark background (line-spectrum of S) took its place. This effect was produced whenever the magnet was excited, and we tried it several times, to make sure of the complete change. After a time, when the magnet, battery, and the coil-power were all weaker, with the magnet on, we obtained a compound of both spectra, the bright lines being seen upon the continuous spectrum in which the bands appeared. When the magnet was taken off, the bright lines disappeared, and the O spectrum alone remained.

SO_3 tube No. 2 examined.

(2) We also tried another SO_3 tube (No. 2) which had been worked for photographic purposes, and was suspected of a carbon impurity. Without the magnet, the spectrum was very like that of the first tube; but when the magnet was excited, the spectrum only brightened, and no bright metallic-looking lines appeared.

---

[15] The tubes generally seem marked Si Fl instead of the ordinary notation Si F. Si Fl_6 is probably, in fact, Si F_4.

Sulphur-tube.

Sulphur-tube. Lighting-up (without heating) described.

(1) A small bent vacuum-tube containing some solid sulphur, excited by the smaller coil, and without being heated, gave a narrow stream of bright blue-green light running straightly through it. With the magnet on, this stream was deflected in the bulbs, and the capillary changed from a blue-green to a distinct rosy tint.

Effect of magnet. Changes in the spectrum.

Without the magnet, the spectrum consisted of four bright bands, with a continuous spectrum between, resembling that of SO_3 tube No. 1. With the magnet on, the spectrum brightened, especially in the yellow and red, which were dull before; and a set of lines appeared upon it (a line or band in the yellow especially showing) which were not seen before. The lines were distinct, but not very bright. The action on the capillary was noticed to be strongest just between the conical points of the armatures; and, in accordance with this, the central part of the spectrum-band in the red and yellow showed an increased brightness.

Effects when one of the bulbs of the tube was heated. Changes in the spectrum under influence of magnet.

(2) One of the bulbs of the tube was then gradually heated with a small gas-flame. The single stream in the heated bulb became somewhat deflected and broken up into a number of smaller streams; and these, when placed under the magnetic influence, had small spark-like threads of light running among them. The capillary, as the tube was heated, and the sulphur rose in it, changed somewhat in tint, and, under the magnetic influence, became of yellow-rose hue. As the heat was applied to the bulb the bands of sulphur gradually appeared in the field of the spectroscope, until at last the band-spectrum of sulphur entirely took the place of the spectrum seen in the cool tube. The magnet being excited, the spectrum changed at once, a set of bright sharp lines (line-spectrum of S) appearing upon a faint and dull image of the band-spectrum.

This effect was constantly repeated upon the magnet being excited. The magnet being taken off, the band-spectrum alone was to be seen.

CHAPTER XV.

EFFECT OF MAGNET ON A CAPILLARY GLASS TUBE.

Capillary portion of a Geissler tube tested in three ways.

The capillary portion of a Geissler tube was cut away from the bulbs, cleaned, and connected by a small vulcanite tube with the gas-pipe in the room conveying coal-gas at ordinary pressure. The flame was small and oval in shape, 8 millims. high, by 4 millims. wide, and burnt quite steadily. (Plate XVII. fig. 13.)

No effect on flame.

(1) The capillary tube was placed between the poles of the excited magnet, almost, but not quite, touching them; no effect at all was produced on the flame.

(2) The tube was placed so that the conical ends of the armatures were allowed to compress the centre of it between them; still no effect was produced on the flame.

(3) The tube was placed so that the straight sides of the armatures compressed it between them; still no effect took place on the flame.

Flame between poles of magnet.

(4) The flame itself was placed between the poles of the magnet. It was slightly drawn towards one pole with an inclination to form the magnetic curve.

Quill glass tubing tested. No effect on the flame.

(5) A piece of quill glass tubing was selected, 5 millims. in diameter and 1 millim. thick, and drawn out to a point, the end of which was snapped off and the tubing connected as before. The flame was 20 millims. high, and 5 millims. across, and somewhat lambent. On being placed (1) between the conical ends and (2) between the flat ends of the armatures, no effect could be seen on the flame.

Effect on taper and spirit-lamp flames.

(6) A small taper-flame was placed between the poles of the magnet: no effect was produced, except that the flame gave a slight "jump" each time the magnet was excited. A spirit-lamp flame was tried with a similar result.

Action of Magnet on a bar of heavy glass.

Heavy glass bar and mounting described.

A piece of heavy yellow-tinted glass was selected, being a bar 10 centimetres in length, and 8 millimetres square. This was mounted in a frame with a Nicol prism at one end, and a double-image prism (next the eye) at the other.

Placed along poles of magnet. Effect of magnet on candle-images.

(1) The glass bar and mounting were placed upon and along the poles of the magnet (in the direction of the magnetic curves), and the double-image prism and Nicol were so adjusted that two images of a candle were seen—the one below bright and normal, the one above, by rotation of the prism, as nearly as possible extinguished (Plate XVII. fig. 4). On exciting the magnet the faint image at once conspicuously brightened, at the same time assuming a slightly green tinge. To get full effect of brightening, it seemed necessary to have good pressure-contact between the battery-wires and the binding-screws.

Effect on using a tourmaline as analyzer.

(2) Using a tourmaline as analyzer in lieu of the double-image prism, the candle-flame was seen alternately brightened and darkened, as the tourmaline was rotated; and when the image was obscured by rotation, excitation of the magnet caused it to brighten strongly. This effect was accompanied by the apparent removal of a dusky red patch or spot, which occupied the centre of the field when the flame was obscured.

Bar placed at right angles to the poles: no effect produced.

(3) The bar of glass and double-image prism being placed between the conical ends of the armatures, but at right angles to, instead of along, the poles, upon excitation of the magnet no effect at all was produced.

(4) The bar and prism being placed in the same position between the flat ends of the armatures, no effect at all was produced.

Slight effect on second experiment.

(4a) Experiment No. 4 was repeated. It was thought that on excitation of the magnet the secondary image slightly brightened; but there was a doubt about it, and the effect (if any) was slight.

Effects produced when a biquartz was introduced.

(5) The apparatus was now changed for one of the following arrangement:—1, a rotating Nicol prism next the eye; 2, the glass bar; 3, a biquartz with the halves horizontal; 4, another Nicol prism. The neutral-passage tint of the biquartz was found to be rather green (from mixture with the yellow of the glass).

Change in colour of the halves.

(i.) Placed *along* the poles of the magnet and the magnet excited, a change of tint was seen in both halves of the biquartz, the slightly purple-reddish tint of the upper half passing into a full purple. Effect not so marked as with the double-image prism.

(ii.) Placed *across* flat ends of the armatures (as in experiment No. 4) no effect was seen.

CHAPTER XVI.

EFFECT OF MAGNET ON WIDE AIR (AURORA) TUBE.

Wide air-tube described.

A large, wide air-tube was tried; it was 14½ inches long by 1 inch in diameter, of the same bore throughout, and with straight platinum electrodes.

Magnet effect when tube placed vertically between conical armatures.

(1) To excite it the larger coil was used. The tube was filled with bright, steady, rosy light, and beautiful stratification, which, as it flickered, seemed to incline to a continuous spiral (Plate X. fig. 8). This stratification was very close and fine, and extended nearly throughout the tube. On excitation of the magnet (the tube having been placed *vertically* between the conical armatures), the glow was condensed into a bright solid line or stream of light at the point which lay directly between the poles. This line or stream expanded into an elongated funnel-shape as it retreated from this centre towards the extremities of the tube, the stratification showing itself more distinctly as the glow of light became less dense (Plate XVIII. fig. 3). The stream of light was driven away at right angles to the poles, and changed from side to side of the tube with the direction of the current.

[With the small coil this tube showed only a flickering stream of light, with very slight indications of stratification.]

Effect when tube placed horizontally between the armatures.

(2) The tube was placed *horizontally* between the conical ends of the armatures. The condensed stratified stream of light flew upwards and downwards (according to direction of current) instead of to the respective sides of the tube.

Tube placed along the poles of the magnet.

(3) The tube was placed along the poles of the magnet. In the interval between these the stream was driven upwards, but at either end sideways, right or left according to whether the pole was N. or S. (Plate XVIII. fig. 4). The result gave a complete spiral of stratified condensed light within the tube.

Note on Stratification.

Stratification in small tubes arranged in series.

The current from the large coil was sent through a set of five small French vacuum-tubes, of equal calibre, containing salts of strontium and calcium, and showing phosphorescent effects. These tubes were arranged in single series; and, from the colour of the glow-discharge, were presumed to contain rarefied air in contact with the salts.

A strong coarse stratification was seen in the central (No. 3) tube. Tubes Nos. 2 and 4 also showed stratification, but in a less degree; while the outside tubes, Nos. 1 and 5, showed no stratification at all. The current was steady, and these effects did not fluctuate.

Effect of Magnet on Plücker (Air-) Tube.

Plücker air-tube. Lighting-up described. Effect of magnet on the positive-pole stream.

(1) A Plücker air-tube was selected of the form shown on Plate V. fig. 1, and was excited by the small coil. The ring was used for the positive pole, the straight electrode for the negative. When lighted up, the tube glowed with a perfectly steady and quiescent light. The negative electrode was surrounded by the usual bright violet glow, extending itself and being gradually lost at a short distance from the wire, while the ring let fall a faint, tubular, salmon-coloured, diffused stream of light, which met the violet glow as it approached the negative pole. The tube was then placed vertically between the poles of the electro-magnet, the armatures being almost in contact with the sides of the tube around the negative pole. On excitation of the magnet, an instantaneous change took place. The stream of light from the positive pole contracted itself, so that it became of a long funnel-shape (the ring forming the mouth of the funnel), while it tapered almost to a point where it met the violet glow.

Effect on negative violet glow.

The stream also became very brilliant (the sides of the tube being left proportionately free from light), and crossing it were a set of bands, or striæ, having a waving or vibratory motion. The whole of the negative violet glow was simultaneously gathered into a brilliant narrow arc, which stretched across between the poles of the magnet. These effects are shown on Plate V. fig. 1. The edges of the arc were remarkably sharp and well defined, and with no surrounding aura or shading off.

Arc of light followed the magnetic curves.

By moving the tube between the armatures it was seen that the arc of light followed the magnetic curves. If the tube was moved upwards, the arc curved towards the zenith, if downwards, contrariwise; and a middle position could be selected, in which the edges of the arc were nearly parallel. Moving the tube a short distance from the pole had the effect of rendering the arc more diffuse, but not of otherwise altering its character.

Direction of the current changed. Effects on glow described.

(2) The direction of the current in the tube was then changed; and, without the magnet, the ring electrode was surrounded by a diffused violet glow; while the straight wire gave forth a faint salmon-coloured stream of light, spreading up to the ring.

Magnet effects described. On negative pole. Rings from positive pole described. Effects on rings of making and breaking contact with magnet. Shape of rings described.

On excitation by the magnet (the positive pole being now placed between the armatures), the violet glow of the negative pole contracted into a compact mass round the ring electrode. At the same time from the positive pole sprang a set of bright saddle-shaped rings, which increased in size as they advanced; and spreading upwards with a rapid but smooth motion towards the negative pole, closely approached to, but never actually came in contact with, the violet glow. The positive end of the tube was otherwise but slightly lighted, and the sudden appearance of this brilliant stream of rings of light was very striking. A single bright ray was also seen running from the positive wire, in a somewhat transverse course, along one side of the tube. When wire-contact with the magnet ceased, so that it was not excited, the rings ran back in succession to the positive pole and disappeared, and by making and breaking contact they were caused to advance and retire at will. They were accompanied by a waving or vibratory motion, and were evidently of the same character as the smaller striæ or bands mentioned as seen when the ring formed the positive pole. The general appearance was that of a hollow cone of light (the base towards the negative pole), composed of brilliant rings with dark spaces between, which appeared and expanded under the magnetic influence, and contracted and disappeared on its removal. The rings did not appear to be flat disks, but were somewhat curved or saddle-shaped. They reminded one much of the diatom *Campylodiscus spiralis*; that is to say, they were apparently flat if looked at from above, but like a figure of 8 when viewed sideways, the peak of the saddle forming a kind of brilliant point or apex.

All this is difficult to describe; but an illustration from a sketch made of the tube is given on Plate XVII. fig. 2.

Negative pole placed vertically on the magnet.

(3) The negative pole (straight electrode) was then placed vertically on one of the poles of the electro-magnet. On excitation, the violet glow was contracted into a small upright brush or column of bright light, with a slight inclination to curvature.

Tube laid horizontally across poles of magnet.

(4) The same Plücker tube was laid horizontally across the poles of the electro-magnet (without armatures), the respective electrodes being above each pole.

Effects produced.

From the negative (straight electrode) pole sprang a dense and compact arc of violet light, in the direction of the magnetic curves, which terminated at the upper circumference of the tube, but which, if prolonged, would have followed the curves to the opposite pole. The stream from the positive pole was very considerably brightened, as in the other experiments, but did not appear in the form of rings or waves. It assumed that of a bright steady continuous glow, which formed round the tube a not perfectly continuous, but distinct and well-marked, spiral. This form of discharge seems connected with the peculiar contour of the rings mentioned in experiment 2. One might, indeed, conjecture the spiral-shaped glow to be a ring of light extended or drawn out towards the negative pole.

Effects like those obtained by Gassiot.

Experiment No. 2 seems in result very like that of Gassiot's with his grand battery and the Royal Institution magnet, the effects (though of course upon a smaller scale) being similar to those obtained by him.

Effect of Magnet on Plücker Tube (Tin Chloride).

Plücker tube (tin chloride). Lighting-up described.

A large Plücker tube was examined, which had a bulb attached at each end, communicating with the central portion by a narrow neck or constriction. On connexion with the small coil, a narrow stream of pale diffused cobalt-blue light ran along the whole tube, from point to point of the electrodes, the positive wire at the same time glowing with an aura of amber-yellow light. (See Plate XVII. fig. 3, where the narrow stream of light is shown by dotted lines.) At the two necks or constrictions the stream of light was perceptibly brightened.

Effects of magnet upon the stream.

When the magnet was connected, the stream in the positive bulb was not much changed, but only slightly bent. In the central partition of the tube and in the negative bulb, the stream of light was broken and split into a number of smaller streams, and at the same time bent or forced against the sides of the tube. (See Plate XVII. fig. 3.)

Peculiar noise within the tube.

In the central partition, the blue streamlets were accompanied by a number of spark-like threads of golden light, which shone out among them as the whole vibrated against the side of the tube; at the same time a peculiar pattering, as of a miniature hail-storm within the tube, made it ring with a slightly metallic tinkle.

The direction of the bending or deflection of the stream was at right angles to the axis of the poles of the magnet, and changed from side to side of the tube as the direction of the current from the coil was varied.

Spectrum described. Without magnet. With the magnet excited.

In the positive bulb the stream, instead of joining the point of the electrode, left this and ran along one side of the whole length of the wire. (See effect, Plate XVII. fig. 3.) The spectroscope was applied to the neck of one of the bulbs where the stream was bright. Without the magnet a faint continuous spectrum, mainly of the blue and green, with very slight traces of the yellow and red, was seen. Upon this, five or six faint but sharp and metallic-looking lines were seen. On the magnet being excited, the continuous spectrum was not changed; but the sharp lines shone out brighter and clearer, one in the blue being especially conspicuous. These lines were measured with a micrometer; and their places being compared with Lecoq de Boisbaudran's "Spectres lumineux," they were easily recognized to be those of tin. On each excitation of the magnet the same brightening of the lines took place.

Effect of Magnet on Tin-Chloride Geissler Tube.

Geissler tube, Sn Cl₄, examined. Glow described. Effect of the magnet.

We then examined a Geissler tube, marked $Sn\ Cl_4$. When first excited by the small coil, the spark passed freely. The glow in the bulbs was of a diffused, light purple tint; the positive electrode had a bright yellow glow around it. The capillary stream was of a sharp green-yellow, at times brightening up to a metallic-looking green. When the magnet was first employed, the tube distinctly and permanently brightened up throughout.

Spiral formed in positive bulb. Glow in tube extinguished.

The negative bulb was not much changed in appearance; but in the positive bulb a curious permanent and steady cloud-like spiral, of a purple colour, made its appearance, and lasted while the tube was under the magnetic influence. (See Plate XVII. fig. 12.) After a short time the tube seemed to lose a great deal of its conducting-power, and to light up in a feeble and intermittent manner, brightening only when the coil was made to work its best. While in this condition, the magnet (which had been previously disconnected) was excited, and at once what moderate glow was still shining in the tube was totally extinguished. At first it was thought some accident might have happened to the conducting-coil wires; but repeated trials satisfied us that the effect was due to the magnetic influence alone. Efforts were made, by looking to the coil and battery, to brighten up the tube as at first, but they quite failed; and it was evident some change had taken place in its conducting qualities. This tube was accidentally broken, so that we had no opportunity to renew the experiments.

Another tin-chloride tube tried. Glow described.

We subsequently tried another tin-chloride tube, purchased of Mr. Browning. This lighted up like the former tube, but brighter. There was an amber glow at the junction of the negative bulb which adjoined the capillary part. This was lost on putting on the magnet. At the same time a perceptible pattering ringing noise was heard in the tube, and metallic-looking threads of light ran through the bulbs.

Spectrum described.

Without the magnet, the spectrum was a continuous faint misty one, with bright lines of tin occasionally flashing up. With the magnet, the tin lines at once shone out bright, strong, and clear upon a black background, the change in effect being very marked.

CHAPTER XVII.

EFFECT OF MAGNET ON BULBED PHOSPHORES-CENT TUBE.

Large phosphorescent bulbed tube. Lighting-up described. Spectrum described. Glow when discharge stopped, described.

Mr. John Browning kindly lent me a large phosphorescent tube with five bulbs, said to be filled with anhydrous sulphurous-acid gas (SO_2). (See Plate XVIII. fig. 1.) This tube lighted up beautifully with the large coil. The connecting tubular parts of it were filled with a bright, beaded, transparent, rosy light; while the bulbs glowed with a more opaque blue-tinted effect. The spectrum of the tubular part was found to agree exactly with the principal bright band seen in a SO_3 Geissler tube. The spectrum of the bulb-glow was a faint green-blue continuous one, with bright bands or lines faintly flashing up at times. When the discharge was stopped, the tube still glowed with a moderately bright, opaque, grey-green light. This glow gradually faded out, always commencing with the bulb forming the negative or violet pole, and so dying out, bulb by bulb, towards the positive pole. The negative-pole bulb at times was, on suddenly stopping the current, hardly lighted at all, the other bulbs being luminous.

Comparison with SO_3 Geissler tube.

(1) We compared the large tube with a SO_3 Geissler tube, by means of a comparison-prism on the slit, with the result before detailed. The Geissler tube, however, showed no after-glow.

Effects in bulbs on lighting-up the tube described. Effects of reversal of the current. After-glow restored by passing of current.

(2) We lighted up the Browning tube with the large coil. The negative bulb was always the least filled with the blue opaque vapour, and the other bulbs increased in vapour-density in the order they approached towards the positive bulb. When the current was reversed, so that the negative and positive glow changed places, the negative bulb still remained transparent, although the positive opaque glow had (presumably) been thrown into it. When the after-glow had quite disappeared in the bulbs, it was again strongly restored, by the passing of the current for a few seconds only through the tube.

Effect of reversal of current on positive-pole glow.

(3) The tube was well excited, and the four bulbs (other than the negative one), upon stopping the current, glowed strongly. The current was then sent through reversed, so as to throw the negative glow for a few seconds into the positive bulb. The after-glow in the positive bulb was at once extinguished. On once more reversing the current, it was only restored after a certain amount of continuance of the positive stream.

Effect of change of current on the three central bulbs.

The time during which the negative glow was thrown into the positive bulb did not appear sufficient to have heated it. After rapidly changing the direction of the current several times and then stopping it, the three central bulbs alone had an after-glow, the two extreme ones having none, being both equally transparent.

Effect of heat on the bulbs. Effect of cooling by ether-spray.

(4) A moderate heat from a spirit-lamp was applied to the centre bulb (*a*) while the current was on; and also (*b*) when this was stopped, and the bulb glowed. In the first case the bulb was found to get more transparent; and in the second case the after-glow disappeared in a proportionately shorter time in the heated bulb than in the others. To test the result of cooling the bulbs, the negative-pole bulb and also the central one were each subjected to the action of ether-spray, and also of ether and water-spray mixed. This was done, (*a*) when the current was passing, and (*b*) when it was stopped and the glow only was in the bulb. The bulbs were cooled until a marked cold effect to the touch was produced. We did not notice any difference in the behaviour of the bulbs so treated as compared with the others, either when the current was passing or in the case of the after-glow.

Negative-pole bulb between the armatures of magnet. Effects on negative and positive glow.

(5) We placed the negative-pole bulb between the conical points of the armatures, and excited the magnet. The negative glow contracted itself into a condensed violet-tinted crescent, in accord with the magnetic curves. The positive glow of the same bulb lost its beaded (stratified) character, and was condensed into a bright stream of light, which latter protruded from the small inner tube and formed a spreading spiral set of cloud-rings within the bulb (see Plate XVIII. fig. 2). The action of the magnet seemed to be exercised in subduing the stratification, condensing the glow into a bright stream of light, and forcing the latter to "tail over" at each extremity of the tubular joints into the bulbs—this effect extending even so far as the second bulb.

When the positive bulb was placed between the poles of the magnet, the glow was simply condensed into a bright stratified stream, which flew to either side of the bulb.

Effect of magnet on glow in bulb No. 4.

(6) *a*. Bulb No. 4 (see Plate XVIII. fig. 1) was placed between the poles of the excited magnet, and the current was passed and then stopped. The glow in that

bulb faded away out of its order, and earlier than in ordinary cases (nearly as soon as No. 2).

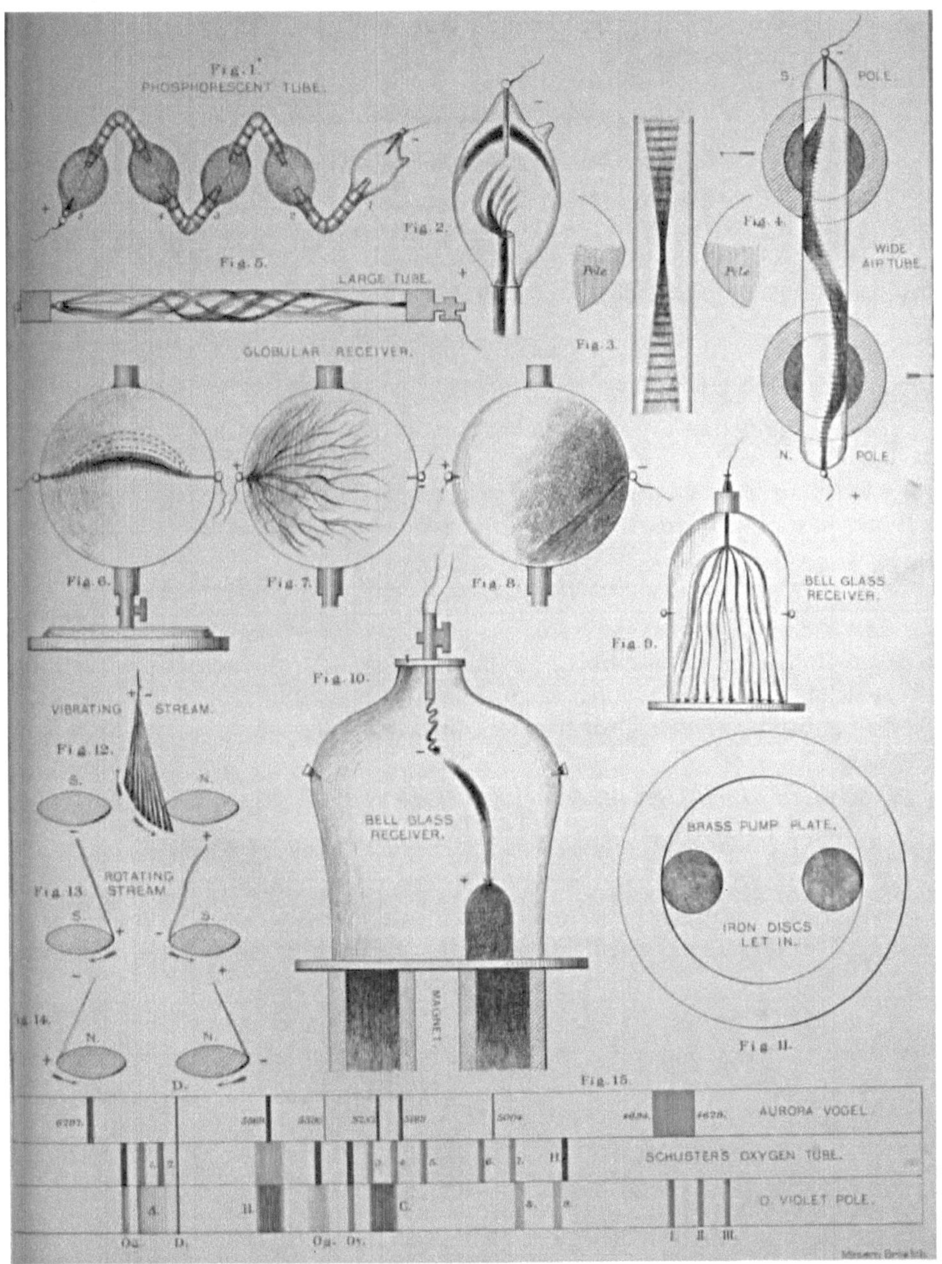

Plate XVIII.

SAME, AND OXYGEN-SPECTRUM

Other bulbs tested in similar manner.

b. The same and other bulbs were tested in a similar manner. In all cases the bulb influenced by the magnet, when the current was stopped, was found perceptibly fainter in after-glow.

Effect of magnet upon the after-glow itself.

c. The tube was arranged with one of the bulbs between the poles of the unexcited magnet; the current was passed and stopped, and the after-glow obtained. The magnet being then quickly excited, the after-glow in the bulb, under its influence, faded out; and the bulb became transparent, perceptibly sooner than under ordinary circumstances. We tried this several times, with the same result in each case.

Mr. Thompson's experiments on action of magnets upon liquid rings.

Note.—In relation to these experiments, it may be mentioned that Mr. S. P. Thompson, of Bristol, is reported to have studied the action of magnetism upon rings of coloured liquid projected through water, and to have observed their retardation and partial destruction in passing through a powerful magnetic field.

Mr. Ladd's explanation of some of the phenomena observed.

Mr. Ladd has suggested to me that some of the phenomena produced indicate a driving of the gas in the direction from the negative to the positive pole—a theory which is supported by the action of the magnet on the bulbs, if this be considered a repulsive one as regards the gas influenced.

Effect of Magnet on small Phosphorescent (powder) Tubes.

Tubes containing phosphorescent powders described.

We examined six vacuum-tubes containing phosphorescent powders, which, upon exposure to sunlight and removal to the dark, or after passing of the electric current over them, continued to glow in the tubes after the exciting cause had ceased. They were of thin glass, and of equal calibre throughout.

One was 6½ inches long and ⅝ inch in diameter, and had no label; the other five were 7½ inches long and ½ inch in diameter, and were labelled respectively:—

Strontium vert,

Strontium jaune,

Calcium violet,

Calcium orange,

Calcium vert-bleuâtre.

Lighting-up of the tubes described. Effect of magnet on ⅝-diameter tube. Spectrum without magnet.

The powders in tubes of this description are said to contain either sulphide of strontium, or calcium, or sulphate of quinine. The first-mentioned tube shone with a white and bright light, and probably contained the latter substance. The general effect of the current on the tubes was similar in all cases. Under a sufficiently strong current, they lighted up with a brilliant, slightly green-white glow; in which, however, by looking sideways, it was possible to detect a delicate rosy tint. Any colours beyond these in the tubes seemed to depend on the powders enclosed in them. When the current was stopped, the powders alone glowed in accordance with the colours mentioned on the labels, the rarefied gas or air in the tubes not giving any after-glow, as in the case of the sulphurous-acid tube. When the ⅝-diameter tube was excited by the small coil, the effect of the magnet was to entirely suppress and extinguish the glow. When this and the other tubes were worked with the larger coil, the spectrum, without the magnet, was bright and continuous, either showing no lines or else very faint traces of them, and, extending through the whole range of colours was brightest in and about the green.

Magnet effect on glow. Same on spectrum.

With the magnet excited, a bright line of pink light was condensed against the upper side of the tube; while the glow in the tube generally became very decidedly fainter, except at the electrodes, which still preserved a certain amount of brilliancy. The spectrum also was much changed. The bright continuous glow became much fainter, and many sharp and fairly bright lines were seen upon it. These lines were, as to character, not easy to recognize. Hydrogen (F) was, however, plainly distinguished; and other lines, which we considered to be N, were common to all the tubes. Some lines were also remarked as being, without the magnet, not so constant.

Tubes examined and compared for spectra.

Calcium orange and calcium violet, compared for spectra, were identical; the two strontium tubes hardly so, but with strontium vert a bright continuous spectrum mainly hid the lines.

Strontium jaune and calcium orange were not alike; strontium vert and calcium violet differed. Calcium orange and calcium vert-bleuâtre were considered alike; but the comparison was not easy, as the calcium vert was bright, and the lines were only seen faintly upon the continuous spectrum.

In order not to shift the powders, the tubes were laid horizontally, and two spectra simultaneously examined across the tubes.

Lighting-up Tubes with One Wire only (Marquis of Salisbury's Observations).

One wire only connected with an electrode.

The vacuum-tubes employed were examined in the usual way, but one wire only was connected with an electrode. The other wire was attached to the end of a glass rod, and circuit was from time to time completed while the tube was before the spectroscope.

The large coil was used. In all cases, with the one wire, the glow was very faint as compared with that of the closed circuit.

Ether vapour.

(1) *Ether Vapour.* — With both wires, in company with the usual bright bands of the carbon spectrum, shading-off towards the violet, the H lines were very sharp and brilliant. With the one wire only, the carbon bands were left faintly shining, with both sides nebulous alike, and with no shading-off towards the violet. (We were not quite sure whether this was not the effect of the reduction of the light.) The H lines, though originally stronger than the carbon bands, quite disappeared from the spectrum.

Coal-gas.

(2) *Coal-gas.* — The same effects were produced; but we thought we could detect very faint traces of the H lines.

Nitrogen.

(3) *Nitrogen.* — The N lines, as well as those of H (also seen in the tube), were much fainter with one wire, but the H lines more so in proportion.

Hydrogen.

(4) *Hydrogen.* — Only a marked reduction in brilliancy of the whole spectrum.

Oxygen, N and H.

(5) *Oxygen.* — An impure tube, showing O (some of the lines hydrocarbon?), N, and H spectra simultaneously. With one wire the O lines still remained fairly bright, the N and H being only faintly seen.

Water-gas.

(6) *Water-gas.* — Same effect.

Turpentine vapour.

(7) *Turpentine Vapour.* — Same effect as ether, but the H lines could be faintly seen.

CHAPTER XVIII.

ACTION OF THE MAGNET ON THE ELECTRIC SPARK.

Apparatus employed.

The magnet was excited with two plates of the large battery, and the larger coil with the other two plates, the action in both cases being strong.

1. A spark from the coil was passed between two platinum wire electrodes, about three centimetres apart.

Spark and aura described.

It consisted centrally of a thin stream of bluish-white light, vividly bright, around which was seen a narrow, uniform, diffuse, yellow-tinted aura, which accompanied the spark in all its movements. The spark always struck across from the extreme points of the electrodes (see Plate XVII. fig. 5).

Effect of magnet upon the aura.

2. On being placed between the conical poles of the excited magnet the bright thread of the spark did not change; but instead of the inconsiderable yellow-tinted aura which accompanied the unmagnetized spark, there now struck out, at right angles to the magnet-poles, a thin rosy-tinted half-disk of aura-like flame. This extended aura ran considerably along each electrode, though the spark proper still struck from the points.

Extended aura described.

The aura was somewhat larger in extent upon one electrode than on the other. In the first case, it sprang from a considerable number of minute illuminated points; on the other electrode, these illuminated points were fewer in number, and the flame was more purple in tint. Reversing the current these effects were reversed. The aura was uniformly thin and disk-like, and the curved edge remarkably true in shape (see Plate XVII. fig. 6).

The lateral direction of the aura was changed when the current was reversed.

Aura not proportionate to length of spark.

3. The aura was found not proportionate to the length of the spark. When the electrodes were approached, so as to very much shorten the spark, the aura still

sprang out to a distance and extent quite out of proportion to the length of the spark. Even when the electrodes were approached so close that the spark was very short indeed, still, under the magnetic influence, a very considerable aura made its appearance.

Effect of working coil-break upon the aura.

4. Upon working the coil-break, it was found that in proportion as the contact screw was drawn apart from the break, so the aura gradually diminished in extent, until at last, by continuing to increase the distance between the screw and the break, a point was reached when thin bright sparks, without any aura, passed. Upon the screw being worked up closer, thicker sparks passed, and the aura again made its appearance. As the aura diminished in size it gradually changed in tint from yellowish rose-pink to purple.

Spark taken in glass bulb.

5. The spark was taken in a glass bulb, the tube in which it was blown being open at both ends, with the same effect as in the open air.

6. A plate of glass was laid on the poles of the magnet, and the spark was passed *along* the poles (in the same direction as the heavy glass was laid in the Faraday experiment). No aura was formed. The points were then moved round, so as to carry the spark at right angles to the poles, and the aura was formed as before.

Aura could be blown away from the spark.

7. The aura, it was found, could be blown away at right angles to the spark. When strongly urged, it assumed the shape of a flickering tongued curtain of flame, flying away in the contrary direction to that from which the current of air proceeded, and again returning to its original shape as the impulse was removed. The spark proper was not influenced (see Plate XVII. fig. 8).

Effect of withdrawing spark from central position between the poles.

8. As the spark was withdrawn from its central position between the poles of the magnet, the convex edge of the aura became gradually less perfect, and assumed a ragged and broken-up appearance, the inequality at times amounting almost to jets or flickering sprays of light. The spark was also slightly curved away from the electrodes (see Plate XVII. fig. 7).

Magnet had no effect upon condensed spark.

9. A condenser of four coated plates was introduced into the circuit, causing a sharp brilliant blue-white spark, apparently divided into streams and with no aura. The magnet had no effect whatever upon this form of spark.

CHAPTER XIX.

THE DISCHARGE IN VACUO IN LARGER VESSELS, AND MAGNETIC EFFECTS THEREON.

A Tate's air-pump was used, and the spark from the larger coil. The exhaustion could not be carried very far.

Globular receiver described. Discharge described.

(1) A globular receiver was used, having brass caps for exhaustion, and platinum wires passing through the opposite sides for electrodes (see Plate XVIII. fig. 6). With partial exhaustion, from the positive electrode proceeded long, sharp, bright, rosy sparks, striking in zigzags across the receiver. From the negative terminal sprang a larger number of bluer and more diffuse streams of light, like spiders' webs; and these were enveloped, for a short distance from the terminal, in a slight misty aura. Both sets played round the sides of the glass as well as across.

Bell-shaped receiver described. Discharge described.

(2) A bell-shaped receiver, with terminals inserted at the sides and one also at the top, was next used (see Plate XVIII. fig. 9). When the side terminals were employed, the effect was much the same as in the last case. When the top terminal was used for one wire (the other wire being connected with the pump-plate) a single stream of bright rosy light ran from the upper terminal to the plate. First striking the central part of the plate, the stream then glided towards one of the lateral terminals, and so to the edge of the receiver. After partly discharging itself by contact with the terminal, the stream as rapidly retreated to the centre of the plate again—this effect being from time to time repeated while the current was passing. The current being reversed, a number of bright, but weaker and more diffused, streams of light had the appearance of shooting from the upper electrode, and of striking upon the plate below; with a tendency to fly off from where they struck, in a similar manner to the single stream before described. Where each stream touched the plate a brilliant point of light appeared, and a strong pattering noise was heard in the receiver.

Bell-shaped receiver without electrodes. Induction discharge described.

(3) Another bell-shaped receiver of similar shape was used. This had no electrodes forming a direct communication with the interior; but, in lieu of these, two wafers of thin sheet brass were cemented, one inside and one outside the glass,

opposite to one another. On connexion being made with the outside wafer, the effects produced by induction were similar to, and very nearly as strong as, those in the cases where direct communication with the interior of the receiver was made.

Long large tube exhausted and illuminated. Spiral form of discharge.

(4) A large tube, 24 inches long and 2 inches in diameter, with ball and point electrodes respectively, was exhausted, and the current passed through it. The effects were similar in most respects to those produced in the globular and bell receivers, but the streams of light assumed a distinctly spiral form in their passage (see Plate XVIII. fig. 5). This tube when placed between the poles of the magnet showed no effect, except a slight condensation of the streams of light towards the sides of the glass.

Globular receiver again used.

(5) The globular receiver first described was again used (the Tate pump having been cleaned and working easier).

Phosphorescent after-glow succeeding the spark.

(*a*) When exhaustion was as good as it could be got, the spark struck across in a single, slightly expanded, stream of rosy light, having a tendency to curve upwards (see Plate XVIII. fig. 6). The electrodes had but little glow round them, only just enough to distinguish the poles apart. When the flow of the stream was interrupted by breaking contact with one terminal, so that sparks passed in succession, we thought we detected a faint blue phosphorescent after-glow succeeding each spark.

Positive wire only attached.

(*b*) The positive wire only was attached to one electrode, the negative being unconnected. A set of faint whity-blue cobweb-looking streams of light spread from the electrode all over the receiver, having a vibratory motion. The spaces between these were dark, and there was no aura—the effect being similar, but not quite so bright and pronounced, as when both wires were attached (see Plate XVIII. fig. 7).

Negative wire only attached.

(*c*) The negative wire only was attached. The cobweb streams were absent, or only shot out very occasionally. The main effect was a straight nebulous stream of violet light, which commenced at the electrode and spread out in a fan-shape towards the lower brass cap of the receiver; while, at the same time, an aura or glow of similar light, but fainter in quality, spread from the electrode over at least one half of the receiver. This aura would no doubt have filled a small flask (see Plate XVIII. fig. 8).

Effect of gradual exhaustion on the discharge.

(*d*) When exhaustion was first commenced, both electrodes of the receiver (both wires being connected) threw out spider-web-like streams, as in Experiment 1, pale blue from the one pole and somewhat rosy from the other.

As the exhaustion progressed the pale-blue streams disappeared, while the rosy flickering ones diminished in quantity and extent until ultimately a single rosy stream of light crossed the receiver as in Experiment 5*a*. Upon admitting the air, these effects took place in an inverse order—the single stream being gradually broken up, and the spider-webs taking its place.

Globular receiver placed on poles of the magnet. Magnet effect.

(*e*) The exhausted globular receiver was placed upon the poles of the excited magnet, with the stream at right angles to them. Looking across the S. pole of the magnet, the negative electrode was on the left hand, and the positive on the right. The effect of the magnet on the stream was apparently to split it up into several; but this appearance must have been due to vibration only, as a revolving mirror showed the stream as single. When the current was reversed, the stream which, without the magnet, was somewhat flickering and vibrating, slightly straightened at the positive pole, and the whole stream became steadier.

Single wires attached.

(*f*) Single wires were successively attached to the negative and positive poles, and the cobweb streamers and glow before described obtained. The magnet was found to have no decided effect on either of these.

Plücker tube placed between poles of magnet with negative wire only attached.

(6) The Plücker air-tube (Plate XVII. fig. 2) was placed between the poles of the magnet, and the negative wire only was connected with the straight electrode. A pale violet glow was seen round this electrode, and another, but rather fainter, glow of a similar description at the ring electrode, the intermediate space being filled with a salmon-coloured light. This violet glow was condensed into an arc by the action of the magnet. Reversing the current, the violet glow still remained at each electrode, and that between the poles of the magnet was still influenced into an arc.

Geissler tube substituted, with similar results.

(7) An air Geissler tube was substituted for the Plücker tube, with very much the same result. Whichever wire was attached, a violet glow appeared at the connected electrode, and a fainter one of the same character at the other; and the magnet influenced both. The connecting salmon-coloured glow was faint.

Globular receiver treated with phosphoric anhydride.

(8*a*) The globular receiver had some phosphoric anhydride shaken into it; and it was then exhausted. The cobweb streamers and violet glow each appeared according to which wire was connected. There was no marked difference between the receiver with the anhydride and without; except that in the former case the streamers and glow were reduced in extent and strength, and were comparatively faint.

Discharge in water-vapour described.

(*b*) The anhydride having been washed out, first with plain and afterwards with distilled water, some drops of the latter were allowed to remain in the receiver. On exhaustion a vapour-cloud was formed, and the discharge passed (both terminals being connected with the coil) through this. The rosy stream of light was formed as usual, but was more flickering and unsteady. As the exhaustion was lessened, the rosy stream disappeared, and the cobweb streams began to fill the receiver. These were, however, not so bright and sharp as in a dry receiver, but were faint and broad; while some diffused and nebulous streams of light, running (slightly bent) from pole to pole, and from ¼ to ⅜ of an inch broad, were inter-mixed with them. When one wire only was connected, the glow and streamers from the electrode were very faint.

Large bell-receiver and plate described. Receiver exhausted and stream of light formed.

(9) A large bell-shaped receiver, 11 × 8 inches, was next used. It was open at the bottom, which was ground as usual; and had a small opening at the top, also carrying a ground edge. A solid brass plate was prepared, ground only round the edge (in order to take the receiver), and in the centre of this brass plate were in-serted two disks of soft iron, corresponding in position and size with the poles of the Ladd electro-magnet (see Plate XVIII. fig. 11). When this plate was placed on the magnet, the poles and disks were in contact; and the disks became N. and S. poles within the receiver. A small brass plate carrying a tap and exhaust-tube, and a binding-screw for attaching an electrode within, closed the receiver at top. The receiver and plate being placed on the magnet-poles, the former was exhausted until the discharge became a rosy slightly-diffused stream of light; with a small unilluminated space between it and the negative pole, where the usual violet glow appeared round the wire.

This stream of light was used for the experiments after detailed. In some cases the conical armatures were placed within the receiver, in others the disks alone were used as the magnetic poles.

Effect on same when magnet excited.

(*a*) With the apparatus arranged as shown on Plate XVIII. fig. 10, and the mag-net excited, a violet glow appeared round the end of the wire which was negative. A small unilluminated space then intervening, the stream ran in a curve between the wire and the armature, which latter was positive. The stream was not steady and had a tendency to rotate; but as this was better observed with the disks only, it is described further on.

Experiments with the conical armature removed. Vibrating stream.

(*b*) The conical armature within the receiver was removed, and the stream allowed to connect with the centre of the pump-plate. When the magnet was excited, the stream was violently projected at right angles to the poles, with a vibrating movement to either side according to the direction of the current. When the wire was positive the movement was towards the left, with a slight inclination towards the N. pole. When the wire was negative the movement was to the right, but in a rather strong curve towards the N. pole. The vibrating motion was very distinct, and gave the appearance of six or seven streams running off at regular intervals (see Plate XVIII. fig. 12).

Rotating wire over S. pole.

(*c*) The wire was next placed over the centre of the disk forming the S. pole. With the wire negative and the pole positive, rotation of the stream was decidedly, but not very strongly, from right to left from the centre of the plate (as the hands of a watch). With the wire positive and the pole negative, rotation was strongly left to right, with a disposition to spiral twist in the stream (see Plate XVIII. fig. 13).

Same over N. pole.

(*d*) The wire was placed over the disk forming the N. pole. With the wire negative and the pole positive, rotation of the stream was left to right. With the wire positive and the pole negative, rotation was right to left (see Plate XVIII. fig. 14).

Stream thrown across the receiver above the magnet-poles.

(*e*) The stream was thrown across the receiver from the lateral binding-screws above, and at right angles to, the disks, and afterwards in the opposite direction, *i. e.* along them. In neither case was there any marked change when the magnet was excited.

(*f*) The conical armatures were placed with the pointed ends upon the disks in the receiver, and the stream thrown above and along them. It diverged—one part running straight across between the electrodes, whilst another stream and some cobwebs ran from each electrode to its nearest pole. The streams and cobwebs flickered a good deal. There was no marked change when the magnet was excited.

Some of Baron Reichenbach's Magnetic Researches tested.

Baron Reichenbach's researches.

In 1846 Dr. W. Gregory published an abstract of Baron Reichenbach's 'Researches on Magnetism and on certain allied subjects, including a supposed new Imponderable.'

Auroræ considered to be magnetic lights. Flames seen by "sensitive" persons.

From a paragraph in this work, it would seem that the Baron considered his observations as tending to an explanation of the Aurora Borealis; and, since it was generally admitted that these phenomena occur within our atmosphere, that there appeared a great probability of Auroræ being visible magnetic lights. The Baron, in the original work, fully describes the Aurora Borealis; and concludes it must be similar in its nature to the flames of light seen streaming from the magnet-poles by Mdlle. Reichel and other sensitive patients of the Baron's. It is unfortunate that these flames were only seen by certain "sensitive" persons. The drawings given of them, too, show no analogy to the magnetic curves.

Magnet tested for such flames.

Having the opportunity of a powerful magnet in that used during our tube-experiments, we made an attempt to detect the Baron's magnetic flames, on or around the poles of our magnet, in a perfectly dark room. Arrangements were made to silently connect and disconnect the battery with the magnet, without the knowledge of any one except the operator. The experiment proved a complete failure; no flames or discharges of light of any kind were to be seen. The observers were five in number, two gentlemen and three ladies, but not one of the party proved "sensitive."

Mr. Brooks's experiments on action of the magnet on a sensitive photographic plate.

Some experiments made by Mr. W. Brooks, and detailed in a paper read by him before the South London Photographic Society, seem to corroborate (to a certain extent) the statements made by the Baron in regard to the influence of the magnet on a sensitive photographic plate.

Remembering, however, how it has been demonstrated that light may be "bottled up" as an actinic source for a considerable period of time, it seems a question whether the images obtained were not due to some such source rather than to any magnetic aura.

SUMMARY OF THE FOREGOING EXPERIMENTS AND THEIR RESULTS.

Summary of the experiments.

Chapter XIV. Action of magnet on glow and spectrum of Geissler gas vacuum-tubes demonstrated.

Chapter XV. Action of magnet on glass capillary tube negatived. Faraday's experiment with heavy-glass bar repeated.

Chapter XVI. Action of magnet on glow in wide air-tube demonstrated. Note on stratification. In Plücker tube, action of magnet on negative pole (arc formed) and positive pole (Gassiot's rings produced) demonstrated. Effects of magnet upon glow and spectrum of tin-chloride vacuum-tubes demonstrated.

Chapter XVII. Effect of magnet upon after-glow in a bulbed phosphorescent tube demonstrated. Effect of magnet upon glow in small phosphorescent (powder) tubes examined. Marquis of Salisbury's experiments (lighting-up with one wire only) tested, and confirmatory results arrived at.

Chapter XVIII. Action of magnet on aura of electric spark demonstrated.

Chapter XIX. Effects of magnet on discharges *in vacuo* in larger vessels demonstrated. Ångström's flask experiment tested; same results not obtained unless one wire only was connected. Experiments demonstrating the action of a magnet on an electric stream, viz. vibration between, and rotation round, poles. Baron Reichenbach's magnetic flames tested without result.

CHAPTER XX.
SOME CONCLUDING REMARKS.

It is usual, in concluding a work on a special subject, to sum up its contents, and to examine the general results arrived at. This, however, it is not easy to do in the present case. The contents of our volume comprise a short history of the Aurora, its qualities and spectrum; and a statement has been given of the several conclusions at which various observers have arrived as to its character and causes. In the present state of our knowledge of the subject, to add an opinion to these might seem to savour of presumption; and the questions involved may perhaps be better treated as still *sub judice*, and as requiring further and fuller evidence before arriving at a verdict. The following observations must therefore be taken rather as further notes and memoranda, than as conclusions. Apart from the spectroscopic questions involved, the oldest and most received theory of the Aurora—that of its being some form of electric discharge in the more rarefied regions of the atmosphere,—seems to hold its own: and if, as is probable, some form of phosphorescence is involved in the discharge, M. Lecoq de Boisbaudran's observations on the brightening of the red line under the influence of cold, and the falling of the yellow-green line within a band of phosphoretted hydrogen, come into play; and a connexion, though slight and imperfect, may be in this respect traced between the discharge and its spectrum. The experiments detailed in Part II. seem to have an important bearing, as showing the very marked effect of the magnet on the rarefied glow, as well as on the spark in air at ordinary pressure. The well-defined arc formed by the aura of the spark, the flickering jets which replace the even edge of the arc when partially withdrawn from the magnetic influence, and the streamers formed when the aura is blown away from the spark (Plate XVII. figs. 6, 7, and 8), are certainly highly suggestive of frequent forms of Auroral discharge; and, but for trial and failure, might lead one to expect results from a comparison of the line air-spectrum with that of the Aurora. The experiments with a wire attached to one electrode only, show how the glow may be affected and varied in colour and character when the discharge is interrupted and incomplete. Differences in electric tension may also considerably vary the character of the discharge.

The influence of the magnet in exciting and brightening the glow and spectrum of one gas, while it depresses and extinguishes the glow and spectrum of another gas in the same tube, suggests an explanation of the observed variation in intensity, and difference in number, of the Aurora-lines. Intensity of lines depending on temperature, and this again on resistance, and it appearing that resistance is influenced by the magnetic action, the same effects of brightening or depressing of the spectrum are probably produced in the Aurora, as in the vacuum-tubes placed between the magnet-poles.

In the Marquis of Salisbury's observations, paraffin-vapour gave C and H lines when connected with both poles of the battery, but C lines only when connected with one pole; and in that case the lines were equally sharp on both sides. These observations (repeated in our experiments) may afford an explanation why the hydrogen-lines are not seen in the Aurora-spectrum; although there can be hardly any doubt that the phenomenon usually takes place in air more or less moist. Professor Ångström's researches on the violet-pole glow are not entirely corroborated by our experiments; and it seems doubtful whether his results in the exhausted flask were not obtained from the negative pole only. One great difficulty in the comparison of the Aurora-spectrum with the violet pole of air-tubes and some other spectra (including oxygen), arises from the presence in the latter of broad bands; and it is difficult to understand how these bands can be aptly compared with the definite, though faint, lines observed by Dr. Vogel and others in the Aurora-spectrum. It must, too, be borne in mind that the conditions under which we may consider the Aurora to obtain, are such as can be only very imperfectly imitated in the laboratory. Auroræ also no doubt differ in density and thickness of layer; and Kirchhoff's observation must be remembered:—"That if thickness of a film of vapour be increased, the lines are increased in intensity, the bright lines more slowly than the fainter; and it may happen that the spectrum appears to be totally changed when the mass of the vapour is altered." Were it possible to test with the spectroscope a cloud or film of gaseous vapour corresponding in some degree in density and thickness with an Auroral discharge, we might perhaps get nearer the truth. Mr. Procter also remarks (as we proved in our magnet experiments):—"That frequently very small traces appropriate to themselves the whole of electrical discharges at low pressures, and completely mask the spectra of any other gases present." The oxygen-spectrum, with its possible variation by the conversion of that gas into the allotropic condition termed ozone, seemed at first to afford a prospect of close relation to the Aurora-spectrum; which, however, disappeared on closer examination. If nitrogen could be modified in some such way as oxygen is converted into ozone, it might perhaps afford another opportunity for investigation; but we have no evidence at present of such a change. The spectrum of nitrogen is usually found singularly distinct and persistent; and, except as varied from band to line by intensity of the discharge, not liable to alteration[16].

Colours of lines are functions of wave-length, subject, however, to the observation that in a weak spectrum the colours lose their intensity. The red line in the Aurora has sometimes been found brighter than the green. It has been suggested that the red and green may be independent spectra; but the variations of tint observed in the capillary of hydrogen and other tubes according to resistance of the current, demonstrate that the varying colours of the Aurora may be connected with the lighting-up of particular parts of the spectrum, and do not necessarily

---

[16] Dr. Schuster has found that while the line-spectrum of lightning is attributable to N, it has also a band-spectrum, which he considers due to O and a slight trace of CO_2 (Phil. Mag. 5th ser. vol. vii. p. 321).

indicate that different gases and spectra are excited.

Absorption may also play an important part in the nature of the Aurora-spectrum (Zöllner's theory that the lines are really spaces between absorption bands). Most gases will give a continuous spectrum under certain circumstances, even at a low pressure.

The question of cosmic dust is inviting, but the facts collated hardly warrant at present its probable connexion with the Aurora.

If Auroræ were composed of incandescent glowing meteors, it would be reasonable to expect to find in the spectrum the lines of iron, a metal constituting so prominently the composition of meteorites. No connexion between the iron and the Aurora-spectra is, however, proved; though it may be suspected. The iron-spectrum, as remarked elsewhere, contains so many lines that some may, as a mere accidental circumstance, closely agree with the Aurora-lines.

The iron-lines are, it may be remarked, as a rule, sharper and finer than the Auroral lines, though it is possible that these characteristics might vary if the spectrum were obtained in a rarefied medium. Tubes with iron terminals are said to evolve a compound gas of H and Fe. I have not had an opportunity to verify this.

It may be added that the comparative faintness of the more refrangible lines of the Aurora-spectrum suggests a feeble resistance to the exciting current, and a low temperature inconsistent with a meteoric theory; and this is not contradicted by the brightness of the red and green lines, if these are due to a phosphorescent origin. Expansion of a line is recognized to be dependent on pressure, and consequently the breadth of the green or red lines might indicate the height of the Aurora; while their brightness or otherwise might also give some idea as to its density. No observations in this direction have, as far as I am aware, been recorded.

As the general result of spectrum work on the Aurora up to the present time, we seem to have quite failed in finding any spectrum which, as to position, intensity, and general character of lines, well coincides with that of the Aurora. Indeed, we may say we do not find any spectrum so nearly allied to portions even of the Aurora-spectrum, as to lead us to conclude that we have discovered the true nature of one spectrum of the Aurora (supposing it to comprise, as some consider, two or more). The whole subject may be characterized as still a scientific mystery—which, however, we may hope some future observers, armed with spectroscopes of large aperture and low dispersion, but with sufficient means of measurement of line positions, and possibly aided by photography, may help to solve. The singular absence of Auroræ has, for some time past, given no opportunity in that direction. May some of my readers be more fortunate in obtaining opportunities of viewing the glorious sky-fires, and assist to unravel so interesting a paradox!

APPENDICES.

APPENDIX A.

REFERENCES TO SOME WORKS AND ESSAYS ON THE AURORA.

(Most of these are cited in the 'Edinburgh Encyclopædia' and the 'Encyclopædia Britannica.')

Musschenbroek, Instit. Phys. c. 41.

'Trai. Phys. et Hist. de l'Aurore Boréale,' par M. de Mairan. Paris, 1754.

Beccaria, 'Dell'Elettricismo Artif. e Nat.' p. 221.

Smith's 'Optics,' p. 69.

D'Alembert's 'Opuscules Mathématiques,' vol. vi. p. 334.

'Philosophical Transactions' as under:—

Vol.	Pages
1716	406
1717	584, 586
1719	1099, 1101, 1104, 1107
1720	21
1721	180, 186
1723	300
1724	175
1726	128, 132, 150
1727	245, 301
1728	453
1729	137
1730	279
1731	53-55
1734	243, 291
1736	241
1740	368
1741	744, 839, 840, 843

1750	319, 345, 346, 499
1751	39, 126
1752	169
1753	85
1762	474, 479
1764	326, 332
1767	108
1769	86, 307
1770	532
1774	128
1781	228
1790	32, 47, 101

'Miscell. Berolinens.' 1710, vol. i. p. 131.

'Comment. Petrop.' tom. i. p. 351, tom. iv. p. 121.

'Acta Petrop.' 1780, vol. iv. p. 1.

'Mem. Acad. Paris,' 1747, pp. 363, 423; 1731; 1751.

'Mem. Acad. Berl.' 1710, vol. i. p. 131; 1747, p. 117.

Schwed. 'Abhandlungen,' 1752, p. 169; 1753, p. 85; 1764, pp. 200, 251.

Bergman, 'Opusc.' vol. v. p. 272.

'Americ. Trans.' vol. i. p. 404.

'Mém. de Mathémat. et Phys.' tom. viii. p. 180.

Rozier, vol. xiii. p. 409; vol. xv. p. 128; vol. xxxiii. p. 153.

Franklin's Works, vol. ii.

Weidler, 'De Aurora Boreale.' 4to.

Nocetus, 'De Iride et Aurora Boreale, cum Notis Boscovisch.' Rome, 1747.

Chiminello, 'Mem. Soc. Ital.' vol. vii. p. 153.

Gilbert's 'Journal,' vol. xv. p. 206; and (particularly) Dr. T. Young's 'Nat. Phil.' vol. i. pp. 687, 716, and vol. ii. p. 488.

Wiedeburg, 'Ueber die Nordlichter.' Jena, 1771.

Hüpsch, 'Untersuchung des Nordlichts.' Cologne, 1778.

Van Swinden, 'Recueil de Mémoires.' Hague, 1784.

Wilke, 'Von den neuesten Erklärungen des Nordlichts,' Schwed. Mus. Wismar, 1783.

Dalton's 'Meteor. Observ.' 1793, pp. 54, 153.

Loomis, 'Sill. Journal,' 2nd series, xxxii. p. 324; xxxiv. p. 34. The same, 3rd series, v. p. 245; B. V. Marsh, 3rd series, xxxi. p. 311.

Oettingen and Vogel, Pogg. Ann. cxlvi. pp. 284, 569.

Galle and Sirks, *ibid.* cxlvi. p. 133; cxlix. p. 112.

Silbermann, 'Comptes Rendus,' lxviii. pp. 1049, 1120, 1140, 1164.

Prof. Fritz, "Geog. Distrib.," Petermann's Mitth., Oct. 1874.

Zehfuss, 'Physikalische Theorie.' Adelman, Frankfort.

'Nature,' iii. pp. 6, 7, 28, 104, 126, 346, 348, 510; iv. pp. 209, 213, 345, 497, 505; x. 211 (Ångström).

'Edinburgh Astronomical Observations,' vol. xiv. 1870-1877.

'English Mechanic,' No. 461 (January 23, 1874), pp. 445-447; and No. 462, pp. 475, 476.

APPENDIX B.

EXTRACTS FROM THE MANUAL AND INSTRUC-TIONS FOR THE (ENGLISH) ARCTIC EXPEDITION, 1875.

Note on Auroral Observations. By Prof. STOKES, *Sec. R.S.*

The frequency of the Aurora in Arctic regions affords peculiar facilities for the study of the general features of the phenomenon, as in case the observer thinks he has perceived any law he will probably soon, and repeatedly, have opportunities of confronting it with observation. The following points are worthy of attention:—

Streamers.—It is well known that, at least as a rule, the streamers are parallel to the dipping-needle, as is inferred from the observation that they form arcs of great circles passing through the magnetic zenith. It has been stated, however, that they have sometimes been seen curved. Should any thing of this kind be noticed, the observer ought to note the circumstances most carefully. He should notice particularly whether it is one and the same streamer that is curved, or whether the curvature is apparent only, and arises from the circumstance that a number of short, straight streamers start from bases so arranged that the luminosity as a whole presents the form of a curved band.

Have the streamers any lateral motion? and if so, is it from right to left or left to right, or sometimes one and sometimes the other, according to the quarter of the heavens in which the streamer is seen, or other circumstances? Again, if there be lateral motion, is it that the individual streamers move sideways, or that fresh streamers arise to one side of the former, or partly the one and partly the other? Do streamers, or does some portion of a system of streamers, appear to have any uniform relation to clouds, as if they sprang from them? Can stars be seen immediately under the base of streamers? Do streamers appear to have any definite relation to mountains? Are they ever seen between the observer and a mountain, so as to appear to be projected on it? This or any other indication of a low origin ought to be most carefully described.

When streamers form a corona, the character of it should be described.

Auroral Arches.—Are arches always perpendicular to the magnetic meridian? If incomplete, do they grow laterally? and if so, in what manner, and towards which side? Do they always move from north (magnetic) to south? and if so, is it by a southerly motion of the individual streamers, or by new streamers springing up to the south of the old ones? What (by estimation, or by reference to known stars) may be the breadth of the arch in different positions in its progress? Do arches appear to be nothing but congeries of streamers, or to have an independent existence? What relations, if any, have they to clouds? and if related, to what kind of clouds are they related?

Pulsations.—Do pulsations travel in any invariable direction? What time do they take to get from one part of the heavens to another? Are they running sheets of continuous light, or fixed patches which become luminous, or more luminous, in rapid succession? and if patches, do these appear to be foreshortened streamers? Are the same patches luminous in successive pulsations?

Sounds (?).—As some have suspected the Aurora to be accompanied by sound, the observer's attention should be directed to this question when an Aurora is seen during a calm. If sound be suspected, the observer should endeavour, by changing his position, brushing off spicules of ice from the neighbourhood of the ears, his whiskers, &c., to ascertain whether it can be referred to the action of such wind as there is on some part of his dress or person. If it should clearly appear that it is not referable to the wind, then the circumstance of its occurrence, its character, its relation (if any) to bursts of light, should be most carefully noted.

These questions are prepared merely to lead the observer to direct his attention to various features of the phenomenon. Answers are not demanded, except in such cases as definite answers can be given; and the observer should keep his attention alive to observe and regard any other features which may appear to be of interest. It is desirable that drawings should be made of remarkable displays.

Observations with Sir William Thomson's electrometer would be very interesting in connexion with the Aurora, especially a comparison of the readings before, during, and after a passage of the Aurora across the zenith.

Spectroscopic Observations. By Prof. G. G. STOKES, *Sec. R.S.*

Spectrum of the Aurora.

The spectrum of the Aurora contains a well-known conspicuous bright line in the yellowish green, which has been accurately observed. There are also other bright lines of greater refrangibility, the determination of the positions of which is more difficult on account of their faintness, and there are also one or more lines in the red, in red auroras.

Advantage should be taken of an unusually bright display to determine the positions of the fainter lines. That of the brightest lines, though well known, should be measured at the same time to control the observations. The character of the lines (*i. e.* whether they are strictly lines, showing images of the apparent breadth of the slit, or narrow bands, sharply defined or shaded-off) should also be stated.

Sometimes a faint gleam of light is seen at night in the sky, the origin of which (supposed from the presence of clouds) is doubtful. A spectroscope of the roughest description may in such cases be usefully employed to determine whether the light is auroral or not, as in the former case the auroral origin is detected by the chief bright line. The observer may thus be led to be on the look-out for a display which otherwise might have been missed.

It has been said, however, that the auroral light does not in all cases exhibit bright lines, but sometimes, at least in the eastern and western arch of the Aurora, shows a continuous spectrum. This statement should be confronted with observation, special care being taken that the auroral light be not confounded with light which, though seen in the same direction, is of a different origin, such, for example, as light from a bank of haze illuminated by the moon.

Sir Edward Sabine once observed an auroral arch to one side (say north) of the ship, which was in darkness. Presently the arch could no longer be seen, but there was a general diffuse light, so that a man at the mast-head could be seen. Later still, the ship was again in darkness, and an auroral arch was seen to the south.

Should any thing of the kind be observed, the whole of the circumstances ought to be carefully noted, and the spectroscope applied to the diffuse light.

Polarization of Light. By W. SPOTTISWOODE, *M.A., LL.D., Treas. R.S.*

It has been suggested that the Aurora, inasmuch as it presents a structural character, may afford traces of polarization. Having reference to the fact that the striæ of the electric discharge in vacuum-tubes present no such feature, the probability of the suggestion may be doubted. But it will still be worth while to put the question to an experimental test.

If traces of polarization be detected, it must not at once be concluded that the light of the Aurora is polarized; for the Aurora may be seen on the background of a sky illuminated by the moon, or by the sun, if not too far below the horizon, and the light from either of these sources is, in general, more or less polarized; therefore, if the light of the Aurora is suspected to be polarized, the polariscope should be directed to an adjacent portion of clear sky, free from Aurora, but illuminated by the moon or sun as nearly as possible similar, and similarly situated to the former portion; and the observer must then judge whether the polarization first observed be merely due to the illumination of the sky.

The presence of polarization is to be determined:—

(1) With a Nicol's prism, by observing the light through it by turning the prism round on its axis, and by examining whether the light appears brightest in some positions and least bright in others. If such be the case, the positions will be found to be at right angles to one another. The direction of "the plane of polarization" will be determined by that of the Nicol at either of these critical positions. The plane of polarization of the light transmitted by a Nicol, is parallel to the longer diagonal of the face; and, accordingly, the plane of polarization, or partial polarization, of the observed light is parallel to the longer diameter of the Nicol when the transmitted light is at its greatest intensity, or to the shorter when it is at its least.

(2) The observation with a double-image prism is similar to that with a Nicol. This instrument, as its name implies, gives the images which would be seen through the Nicol in two rectangular positions, both at once, so that they can be directly compared; and when in observing polarized light the instrument is turned

so that one image is at a maximum, the other is simultaneously at a minimum. Both these methods of observation, (1) and (2), are especially suitable for faint light; because in such a case the eye is better able to appreciate differences of intensity than differences of colour.

(3) The observation with a biquartz differs from (1) only by holding a biquartz (a right-handed and a left-handed quartz cemented side by side) at a convenient distance beyond the Nicol, and by observing whether colour is or is not produced. If the Nicol be so turned that the two parts of the biquartz give the same colour (choose the neutral tint, *teint de passage*, rather than the yellow), we can detect a change in the position of the plane of polarization by a change in colour, one half verging towards red, the other towards blue. This observation is obviously applicable to a change in the plane, either at different parts of the phenomenon at the same time, or at the same parts at different times.

(4) We may use a Savart's polariscope, which shows a series of coloured bands in the field of view. For two positions at right angles to one another corresponding to the two critical positions of a Nicol, these bands are most strongly developed; for two positions midway between the former the bands vanish. In the instruments here furnished, the plane of polarization of the observed light will be parallel to the bands when the central one is light, perpendicular to them when the central band is dark.

Instructions in the use of the Spectroscopes supplied to the Arctic Expedition. By J. Norman Lockyer, *F.R.S.*

Spectroscopic Work.

Scales prepared on Mr. Capron's plan, together with forms for recording positions, also accompany the instrument.

A. In using these, carefully insert the principal solar lines in their places on the forms, as taken from a fine slit, and keep copies of this scale for use. If the slit opens *only on one* side, note on scale in which direction the lines widen out, whether towards red or violet. Also fill up some of these forms with gas and other spectra, as taken at leisure *with the same instrument* and scale.

When observing, close the slit (after first wide opening it) as much as light will permit, and then with pen or pencil record the lines as seen upon the micrometer-scale on the corresponding part of the form, and note *at once relative intensities* with Greek letters, α, β, &c. (or numbers).

Reduce at leisure line-places on scale to wave-lengths, and note as to each line the *probable limits of instrumental error.* **B.**

In case the auroral spectrum is so faint that the needle-point or micrometer-scale is invisible, half of the field of view may be covered with tinfoil, with a perfectly straight smooth edge running along the diameter of the field, in perfect focus, and parallel to the lines of the spectra. The reading-screw being set to 10, the bending-screw should then be adjusted so that the green line of the Aurora is just eclipsed behind the blackened edge of the tinfoil. A similar eclipse of other lines will give their positions.

In this instrument the reference-prism is brought into action by turning the slipping piece to which is fixed the two terminals. Care should be taken that the prism itself is adjusted before commencing observations, as it may be shaken out of position on the voyage. The tubes provided for the reference-spectra may be either fastened to the terminals or arranged in some other manner. The air-spectrum may also be used as a reference-spectrum. To get this, two wires should be screwed into the insulators, their ends being at such a distance apart and in such a position that the spectrum is well seen.

General Observations regarding the Spectrum of the Aurora[17].

C. Note appearance, colour, &c. of *arc*, *streamers*, *corona*, and *patches* of light.

Get compass positions of principal features, and *note any change of magnetic intensity*. If corona forms, take its position and apparent height.

Look out for *phosphorescence* of Aurora and adjacent clouds. Listen for reported sounds. Note any peculiarity of cloud scenery, prior to or pending the Aurora.

Sketch principal features of the display, and indicate on this sketch the parts spectroscopically examined.

Examine line in *red* specially in reference to its assumed connexion with *telluric* lines (little **a** group), and note *as to its brightening in sympathy with any of the other lines*.

Examine line in yellow-green (Ångström's) as to *brightness, width*, and *sharpness (or nebulosity)* at the edges. Notice as to a peculiar *flickering* in this line sometimes seen; note also whether this line is *brighter* (or the reverse) *with a fall of temperature*. Note *ozone* papers at the time of Aurora.

Note whether the Auroræ can by their spectra be classed into distinct types or forms, and examine for *different spectra* as under:—

α. The auroral *glow*, pure and simple.

β. The white arc.

γ. The streamers and corona.

δ. Any phosphorescent or other patches of light, or light cloud in or near the Auroræ. **D.**

The information collected together in the 'Manual' should be carefully consulted, and the line of observations suggested by Ångström's later work followed out. To do this, not only record the positions of any features you may observe in the spectrum, but endeavour to determine, if any, and if so which, of the features vary together. Compare, for instance, the two spectra of nitrogen in the Geissler

[17] *In these observations some suggestions made by Mr. Capron have been incorporated.*

[This was Mr. Lockyer's note. In point of fact, the Author was responsible for the verbatim paragraphs comprised between the letters **A** and **B**, and **C** and **D**, in the instructions as now reprinted.]

tube supplied, by observing first the narrow and then the wider parts of the tube. It will be seen that the difference in colour and spectrum results simply from an addition to the spectrum in the shape of a series of channelled spaces in the more refrangible end in the case of the spectrum of the narrow portion.

Try to determine whether the difference between red and green Auroras may arise from such a cause as this, and which class has the simpler spectrum.

See whether indications of great auroral activity are associated with the widening or increased brilliancy of any of the auroral lines.

Remember that if auroral displays are due to gaseous particles thrown into vibration of electric disturbance, increased electric tension may either (1) dissociate those particles and thus give rise to a new spectrum, the one previously observed becoming dimmer; or (2) throw the particles into more intense vibration without dissociation, and thus give rise to new lines, those previously observed becoming brighter.

Careful records of auroral phenomena from both ships may enable the height of some, observed from both, to be determined. It will be very important that those the heights of which are determined by such means should be carefully observed by the spectroscope, in order to observe whether certain characteristics of the spectrum can be associated with the height of the Aurora.

APPENDIX C.

EXTRACTS FROM PARLIAMENTARY BLUE BOOK, CONTAINING THE "RESULTS DERIVED FROM THE ARCTIC EXPEDITION 1875-76." (EYRE AND SPOTTISWOODE, 1878.)

Auroras observed 1875-1876, at Floebery Beach and Discovery Bay.

By Lieutenant A. C. PARR, *R.N.*

Though the auroral glow was often present, and served in some degree to lighten the darkness of the sky during the long winter, when the moon was absent, the actual appearances of the Aurora itself were few, and the nimbus worthy of any particular remark extremely small. Those which were stationary assumed the form of low arches, with streamers flashing up to them from the horizon, and usually to the eastward. But the more common form was for an arch to appear low down in some part of the sky where the glow was brightest; at first it was very faint and narrow, but as it rose gradually in the heavens it would increase both in size and intensity, till on arriving near the zenith, with its ends extending nearly to the horizon, it would be about the breadth of three or four rainbows, and its colour that of white fleecy clouds lit up by the rays of the full moon. On reaching this point, however, its course was nearly run; for after appearing to remain stationary, as little white gaps would suddenly rend the arch asunder, the portions thus detached seemed to roll together and concentrate all their brightness in the smaller space, and then gradually fade away and become extinct. Sometimes a very pale green would show itself in the more luminous patches, and once or twice there was a slight suspicion of red; but never was the whole sky illuminated by streams running in all directions, and forming coronæ, while these colours varied every moment.

When instead of the arch rising up from the horizon a streamer appeared, its origin was in the north. From the northern horizon it would stretch out towards the zenith, passing nearly overhead, and reaching to within a few degrees of the land to the south. In appearance they would be the same as the arches, but sometimes a second would grow out of the first, and on one occasion three were visible at the same time. They had lateral motion either from east to west, or west to east, but there was no flashing to brighten them, and they gradually faded away.

The time at which Auroras usually occurred was between 9 P.M. and midnight, the last display being on February 19th, commencing at 11 P.M. It was a beautifully clear night, without mist or haze of any description, and small stars visible close down to the horizon. At the above-named hour two arches made their appearance, and remained stationary; the lower one was the brighter, being of a pale green colour, its centre bearing E.S.E. (true), and having an altitude of about 5°, with a breadth of about twice that of a rainbow. The second arch was concentric with the first, and about 7° above it, but rather broader and fainter. These arches maintained their altitude, the upper one at about the same intensity, but that of the lower one varied considerably. It would gradually lighten up, then send flashes to the upper one, then break up and fade away; before, however, it had quite disappeared, flashes would come up to it from the horizon which seemed to endue it with new life, for the arch would be reformed, brighten up, and the same performance would be again repeated. This occurred three or four times in the course of three quarters of an hour; but the flashes from the horizon never extended beyond the lower arch, and those from the lower never went beyond the upper. During this display the citron-line was obtained very clearly with the spectroscope, but no other lines were visible.

On six or seven occasions Auroras were visible at the same time on board both the 'Alert' and 'Discovery;' but the absence of characteristic features makes it impossible to determine whether they were the same display, or merely two distinct ones which happened to occur at the same time. But as by far the larger number of those recorded in the one ship were not visible at the other, it was certainly only under exceptional conditions that they could be simultaneously observed at both stations, if, indeed, they ever were. Auroras seemed to appear indifferently both when there was wind and when it was calm, with either a high or low barometer, and seemed quite unconnected with the temperature, although on an occasion the thermometer was observed to fall 3° during the display, and to rise 2° almost immediately afterwards. But it was never seen illuminating the edges of clouds, as we saw it on the passage home, nor playing about the outline of the land, and never was there the slightest suspicion of sound being produced by it.

The opportunities for observing the spectrum of the Aurora in this position have been most unsatisfactory, as the displays were small in number and deficient in brilliancy.

The form they generally assumed was to rise like an arch from a portion of the horizon where there was a luminous glow, at first very faint, but gradually increasing in brilliancy till near the zenith, where it would remain stationary for a short time and then break up and disappear. Sometimes they would rise up as streamers, but only occasionally was more than one visible at a time, and they lasted for such a short time, that even if they had been bright it would have been very difficult to make satisfactory observations.

Very few showed any signs of colour, and those only the slightest tinge. Nearly all that were observed gave the citron-line with the small pocket spectroscope with more or less distinctness, though no signs of any other lines were ever seen; but on only two occasions was it bright enough to get the line with Nury's spectroscope, and then only for such a short time that a satisfactory measure could not be obtained.

Then follows a descriptive list of the Auroræ seen, from which I have selected three of the finest, viz. January 2nd, February 14th, and February 19th, 1876.

January 2nd, 1876. Lieut. Parr. *Floeberg Beach.* —9 P.M. Streams of Aurora. Stars shining brightly.

Register. *Discovery Bay.* —9 P.M. Observed an Aurora like a pale band of light in the form of an arch whose centre was on the true meridian and 15° from the zenith. It shortly afterwards broke up into feathered edges, their direction being a little to the eastward of the zenith. The arch grew fainter, and shifted to the eastward of the meridian four points; the left extremity of the arch faded away, and the right assumed the shape of the folds of a curtain doubled over. The weather was clear and calm. The display lasted upwards of 30 minutes.

A spectroscope, one of Browning's 8-in. direct-vision, was directed towards the Aurora, but the light was not sufficient to give any spectrum.

The temperature was -39°. Barometer 29·56 inches. No wind. Clouds stratus 2. Eight meteors were observed during the time the Aurora was visible.

February 14th. Register. *Discovery Bay.*—At 2 A.M. a faint Aurora passing across the heavens from S.E. to S.W. was observed, like an arch of a pale colour. It lasted only a short time, and was very indistinct. Temperature -47°. Barometer 30·44 inches. No wind or clouds.

Lieut. Aldrich. *Floeberg Beach.* —2 A.M. A faint Aurora towards the S.W. Weather calm. Cumulus-stratus clouds 3. Temperature -46°. 8 P.M. Faint flashes of Aurora in the E. and S.W.

Lieut. Aldrich and Lieut. Parr. *Floeberg Beach.* —11.50 P.M. A moderately bright arch of Aurora extended from due N. to about S.S.W., where it terminated close down to the horizon in a crook turned to the eastward. In a few moments a streamer flashed from the end of the crook parallel to the first and right across the heavens, its edges being quite sharp and parallel to each other. A third streamer shot up a minute afterwards, but did not extend more than 80° upwards. The streamers were visible for a very short time, the first remaining longest. The second-named arch gradually faded away till within a few degrees of the S.S.W. horizon, and (still being a continuation of the crook) bent round to the eastward, and towards the horizon, going on to what was left of the stump of the third arc. A lateral motion to the eastward now began, the whole body gradually turning round until it disappeared about due south. Stars were visible through it at its brightest, but not very distinctly. This is the most intense and variegated Aurora we have experienced, but scarcely any colours were to be seen. Temperature -51°. Barometer 30·43 inches, stationary. Calm weather. Clouds cumulus 1. Preceded and followed by calm weather.

Meteorological Register. *Discovery Bay.* —9.15 P.M. An Aurora was observed to the southward, spreading out like a fan in separate ways. It was faint. A few cirro-stratus clouds were visible, apparently between the observer and the Aurora. It lasted about 40 minutes, and then gradually faded away. Temperature -47°. Barometer 30·51 inches, stationary. No wind. Clouds cirro-stratus 4.

February 19th. Meteorological Report. *Discovery Bay.*—9.45 P.M. An Aurora like a fluted arch, with rays flashing towards the Pole, was observed spanning the hills from the south to the east. The direction of the lines of light from all parts of the arch was towards the zenith. Above the arch a pale band of colour appeared, like a secondary arch above the other. It appeared very much as if it was caused by the reflected light of the Aurora. The Aurora was bright for a few seconds, and then gradually died away. It lasted altogether about 30 minutes. The centre of the arch bore S.E., having an altitude of about 30°. The secondary arch was about 15° above the former. Both arches were of a pale light colour, the upper one very faint. Temperature -34°. Barometer 29·87 inches, rising rapidly. Weather calm. Misty. No clouds.

Lieut. Parr. *Floeberg Beach.*—An Aurora appeared shortly after 11 P.M., consisting of bright arch, whose centre bore about E.S.E., and had an altitude of about 5°, with a second broader and fainter arch about 7° above the first. These arches maintained their altitudes, the upper one at about the same intensity, but that of the lower one varied considerably. It would gradually brighten up, then send streamers up to the second, then break up into light patches, and gradually fade away. This happened three or four times during the 40 minutes that the display lasted. At times streamers would come up from the horizon to the lower arch, for it was a splendidly clear night, and seemed to brighten it up, but none of them extended beyond it. Neither did the streamers from the lower arch extend beyond the upper one. It was slightly green in colour when brightest, and the citron-line was well defined, but no others were visible. Temperature -46°. Barometer 29·95 inches, steady. Weather calm. Cumulus clouds 4. Misty.

TABLE of DATES when AURORAS were observed by the ARCTIC EXPEDITION, 1875-76.

Date.			H.M.S. 'Alert,' Floeberg Beach.	H.M.S. 'Discovery,' Discovery Bay.
1875,	October	25	11.45 P.M. Faint.	Cloudy.
"	"	26	10 P.M. Very faint.	10 P.M.
"	"	30	Sky obscured. Faint.	Ditto.
"	November	1	Ditto. Faint, but well marked.	Ditto.
"	"	2	9 to 10 P.M. Arches and streamers.	A few clouds.
"	"	21	Ditto. Bright streamer.	9 to 10 P.M. and 10 to 11 P.M.
"	"	22	2 P.M. and 8 P.M. Slight, red.	Clear sky.
"	"	25	9.30 A.M. Character not recorded.	Ditto.
"	"	26	10 A.M. Stream of light.	A few clouds.
"	"	26	Cloudy to 10 P.M., bright afterwards.	10 P.M.
"	"	27	Midnight. Slight.	11.40 P.M.

"	"	28	1 A.M. Bright streak.	Clear sky.
"	"	29	Cloudy, brighter at 11 A.M. Faint glow.	9.30 A.M.
"	"	30	A few clouds. Very faint.	4.30 A.M.
"	"	30	5 P.M., 8 P.M., and 10 P.M. Flashes.	5 P.M.
"	December	2	Evening. Streamers.	Clear sky.
"	"	3	1 A.M. Flashes.	Ditto.
"	"	3	Bright sky. Faint Aurora.	2.30 P.M.
"	"	16	10 P.M. Slight; showed citron-line.	11 P.M.
"	"	19	3 P.M. to 5 P.M., faint; and 9 to 10 P.M., moderately bright arc.	Very clear sky.
"	"	22	10 P.M. Slight.	Ditto.
"	"	23	6 P.M. Ditto.	Ditto.
"	"	24	Misty, a few stars visible. Arch.	9 A.M.
"	"	26	Very bright sky. Faint.	6 P.M.
"	"	29	Ditto. Very faint.	6.15 P.M.
"	"	31	4 P.M. Same.	Sky obscured.
1876,	January	1	5 P.M. and 11 P.M. Slight.	A few clouds.
"	"	2	9 P.M. Described and figured.	9 P.M.
"	"	17	Very bright sky. Very faint streamers.	9.25 A.M.
"	"	18	9.45 P.M. and 10.5 P.M. Character not recorded.	10.15 P.M.
"	"	19	Very bright sky. Faint.	9.45 P.M.
"	"	20	2 A.M. Slight.	2.30 A.M.
"	"	23	7.55 A.M. and 2 P.M. Slight.	8.45 P.M.
"	"	24	Bright sky. Slight flash.	2 A.M.
"	"	24	5 P.M. and 11.15 P.M. Faint Aurora.	Very clear sky.
"	"	27	2 A.M. to 3.45 A.M. Faint.	1 A.M. to 4 A.M.
"	"	27	Very bright sky. Faint double arch.	8.30 P.M.
"	"	28	6 P.M. and 7 to 9 P.M. Faint flashes.	7.20 P.M.
"	"	30	8 P.M. Streak.	7.50 to 9 P.M.
"	"	31	8.30 A.M. and 7.30 P.M. Very faint.	8.25 A.M., 5.30 P.M.
"	February	3	10 P.M. Slight flash.	Very clear sky.
"	"	11	Sky obscured. Very faint.	11 P.M.
"	"	13	11 P.M. Flashes.	Clear sky.

″	″	14	2 A.M., 9.15 to 10 P.M. Described and figured.	2 A.M. and 11.50 P.M.
″	″	19	9.45 P.M.	11 P.M.
″	″	20	2 A.M. Very faint.	2.30 A.M.
″	″	22	2 A.M. Character not recorded.	Very clear sky.
″	″	24	Bright sky. Very faint.	Midnight.
″	″	26	10 P.M. and 11 P.M. Faint flashes.	Sky obscured.

I have added to the above Table the character of the Aurora in each instance as taken from the fuller descriptions given.—J. R. C.

Auroras and Magnetic Disturbances.

The appearances of Auroras and the synchronous movements of the declinometer-magnet were subjects of special observation during the stay of the 'Alert' and 'Discovery' at their winter-quarters. The Table on page 187 gives the dates and hours when Auroras were visible. On all occasions they were observed to be faint, with none of those brilliant manifestations which are described by our own officers as seen at Point Barrow, and by the Austro-Hungarian Expedition at Franz-Josef Land, where the magnetical instruments were so sensibly disturbed.

These phenomena were not observed either in the 'Alert' or the 'Discovery,' especially no connexion between magnetical disturbances and the appearances of Auroras could be traced.

This is quite in accordance with the remarks of previous observers within the region comprehended between the meridians of 60° and 90° west, and north of the parallel of 73° north. For example:—

In the Phil. Trans. 1826, Part IV. p. 76, Capt. Parry and Lieut. Foster remark, in the discussion of their magnetical observations at Port Bowen:—"As far, however, as our own observations extended, we have reason to believe that on no occasion were the needles in the slightest degree affected by Aurora, meteors, or any other perceptible atmospheric phenomenon."

Again, in the Smithsonian Contributions, vol. x., 1858, Mr. A. Schott, in his discussion of Dr. Kane's observations at Van Rensselaer Harbour, in 1854, remarks—"In conformity with the supposed periodicity of this phenomenon as recognized by Professor Olmstead, no brilliant and complete Auroras have been seen; with an exception of very few, they may all be placed in his fourth class, to which the most simple forms of appearances have been referred." The following statement is given in the same page as a footnote:—"The processes have no apparent connexion with the magnetic dip, and in *no* case did the needle of our unifilar indicate disturbance."

The following description of the Aurora observed on 21st November, 1875, is given by Commander Markham and Lieut. Giffard, in their abstract of observations at Floeberg Beach:—

"Between 10 and 11 P.M. bright broad streamers of the Aurora appeared 10° or 15° above the north horizon, stretching through the zenith, and terminating in an irregular curve about 25° above the south horizon, bearing S.S.W. During the Aurora's greatest brilliancy the magnet was observed during five minutes to be undisturbed."

[*Note.*—I applied for a loan of the lithographic stones to enable me to give copies of the three diagrams of Auroræ referred to in the Arctic "Results;" but the Lords Commissioners of H.M. Treasury refused this, except on the terms of my paying one third of the original cost of production of such diagrams. I did not think it worth while to accept these conditions. Only one of the drawings has any special interest; and this is a "curtain" Aurora, similar to that figured on Plate II. of this work.—J. R. C.]

APPENDIX D.

THE AURORA AND OZONE.

Aurora and ozone. Dr. Allnatt's notes and conclusions deduced therefrom.

While Part I. was in the press, Dr. Allnatt, formerly of Frant, and for many years the well-known meteorological contributor to 'The Times' newspaper, kindly placed at my disposal his large series of notes. Upon an examination of these we came to the following conclusions:—

1. That Auroral periods are also periods of comparative abundance of ozone.

2. That instances are by no means wanting in which an abnormal development of ozone appears to be coincident with the manifestation of an Aurora.

Year 1870 remarkable for sun-spots, auroræ, and ozone.

In reference to the first point, it is found, as the result of an examination of Dr. Allnatt's notes, that particular years and months are notable at once for Auroræ and for ozone in abundance. 1870 was one of these years, and was specially recorded by Dr. Allnatt, in his 'Summary for the Year,' as remarkable for sun-spots, Auroræ, and ozone.

Particulars of some of the monthly records.

The month of February in that year was marked by intense cold and brilliant Auroræ. Atmospheric electricity was feeble, but ozone was, throughout the month, well developed; and there was no tangible period of antozone.

In the month of April of the same year, eight days consecutively (19th to 26th) were marked for ozone 10, the maximum of Dr. Allnatt's scale.

In May of the same year there were magnificent Auroræ, and atmospheric electricity was intense. Ozone was scanty; but this was accounted for by the wind being generally E.N.E., ozone being mostly developed with a W. or S.W. wind, and a moist state of the atmosphere.

In August 1870 the unusually large number of 22 days were recorded for a maximum of ozone.

September 1870 was hardly less remarkable, with 19 days of maximum. It was recorded that there were splendid Auroræ during this month, and the solar spots were very large.

October 1870 had 20 days of maximum ozone, and November had several fine Auroræ and maxima of ozone noted. In fact, nearly every month in that year was referred to by Dr. Allnatt for displays of Aurora (of both Arctic and Antarctic forms) and for a development of ozone very considerably above the average.

Year 1871.

The year 1871 had more or less of the same character. In the month of October of that year, fine Auroræ were prevalent, and ozone was registered as at its maximum during 22 days.

There seems reason to conclude that if a systematic comparison of annual or other periods of Aurora and ozone development were made, it would result in disclosing a connexion (probably an intimate one) between the two phenomena.

Instances showing a connexion between a specific Aurora and an ozone maximum.

With reference to the second point, the following (among other) instances may be quoted, for the purpose of showing a connexion between a specific Aurora and an ozone maximum.

The Aurora of 24th September, 1870, was splendid and universal, being seen in Europe, Asia, Africa, America, and Australia. Ozone reached, on the morning of the 24th, 8 of the scale (the scale running from 1 to 10), and, on the morning of the 25th, 10, the maximum.

In October 1870 there were grand displays on the 14th, 20th, 22nd, 24th, and 25th, and ozone was correspondingly abundant, as is seen by the following Table: —

Date.			Aurora.	Ozone.	
1870,	October	14th.	Aurora.	8	
"	"	20th.	Aurora.	10	The display of the 24th was accompanied by the formation of a corona, and that of the 25th was splendidly seen in Edinburgh.
"	"	21st.	None seen.	5	
"	"	22nd.	Aurora.	10	
"	"	23rd.	None seen.	8	
"	"	24th.	Aurora.	10	
"	"	25th.	Aurora.	8	

The foregoing figures somewhat point to the conclusion that ozone quantity rises and falls coincidently with the Aurora displays.

The following seems, however, a case still more strongly in point.

Date.			Wind.	Aurora.	Ozone.
1871,	January	25th.	E.S.E.	None seen.	0
"	"	26th.	N.N.W.	None seen.	2
"	"	27th.	E.S.E.	Aurora at night in N. and S. horizons.	10

"	"	28th.	E.	None seen.	8
"	"	29th.	S.E.	None seen.	2

It is curious, in examining the above Table, to note how the ozone rose, notwithstanding an east wind, from 0 on the 25th, and 2 on the 26th, to 10 on the 27th, when the Aurora appeared, and 8 on the 28th, when it might have lingered; and how it again descended to 2 on the 29th.

The case of the Aurora of 6th of October, 1869, when a broad belt of Aurora was in the north, is also an illustrative one, as will be seen by the following data:—

Date.			Wind.	Ozone.	Aurora.
1869,	October	5th.	S.S.W.	1	—
"	"	6th.	S.S.E.	5	Aurora.
"	"	7th.	S.S.W.	10	—
"	"	8th.	S.	10	—
"	"	9th.	S.E.	5	—

The Aurora of the night of the 6th was here represented by the ozone-paper of the morning of the 7th with a maximum of 10, which lasted till the 8th.

[It should be borne in mind, in examining these Tables, that the Aurora is of the night of the given date, while the ozone-papers are taken and recorded in the morning of the date quoted.]

Other instances.

We will now take instances where the ozone has not reached its maximum; but even in these cases a certain amount of rise and fall of the ozone development towards and from the Aurora is traceable.

Date.			Wind.	Ozone.	Aurora.
1871,	April	8th	S.S.E.	5	Aurora on 9th, but wind E. and unfavourable to ozone.
"	"	9th	S.S.E.	8	
"	"	10th	S.E.	5	
1871,	November	9th	N.	5	Aurora on all three nights.
"	"	10th	N.W.	8	
"	"	11th	N.	5	
1872,	February	3rd	S.W.	4	Aurora on night of the 4th represented by ozone-paper of morning of the 5th.
"	"	4th	S.S.W.	5	
"	"	5th	S.W.	8	
"	"	6th	S.W.	5	

Other cases are, we are bound to say, found, in which ozone was either not remarkable for quantity, or positively fell during the Aurora, as, for instance, this:—

Date.			Wind.	Ozone.	Aurora.
1874,	March	16th	W.N.W.	6	
"	"	17th	S.W.	6	Aurora on the 18th represented by test-paper of the 19th with only two degrees of discoloration.
"	"	18th	W.	5	
"	"	19th	S.S.W.	2	

It is, however, possible that such instances may be accounted for, either by some reaction on the test-papers after they have been coloured, or by some accidental antagonistic circumstance affecting the tests. The following is a case well illustrating this: —

Date.			Wind.	Ozone.	Aurora.
1874,	January	31st	N.N.W.	6	
"	February	1st	N.W.	8	There was an Aurora on the night of the 2nd represented by the ozone-paper (4 only) on the morning of the 3rd.
"	"	2nd	N.W.	2	
"	"	3rd	N.N.W.	4	
"	"	4th	E.N.E.	8	

This instance would seem strongly opposed to the theory of a connexion between Aurora and ozone but for the fact that on the 2nd, when the Aurora was seen at night, and on other days in the same month, Dr. Allnatt has recorded a strong wave of antozone to have swept over the whole of England, and blanched the ozone-papers, however deep their coloration might have previously been. Indeed, it is easy to understand that some antozonic influence may, at times, disturb the evidence of the test-papers, even in so elevated and apparently pure an atmosphere as that of Frant.

It may not be considered that the foregoing instances are enough to establish a case of ozone=Aurora; but there seems, at least, sufficient to base a requisition for further inquiry upon.

It would, too, be interesting to investigate whether Auroræ and ozone development are respectively localized. Mr. Ingall's fine Aurora, seen at Champion Hill, S.E., July 18th, 1874 (*antè*, pp. 22 and 23), was not observed at Frant, and the ozonoscopes there were described as blanched by antozone.

APPENDIX E.

INQUIRIES INTO THE SPECTRUM OF THE AURORA.

By H. C. Vogel.[18]

The frequent appearance of the Aurora in the past winter, as well as this spring, has given me opportunity to institute exact inquiries into the spectrum of the Aurora. It is known that the nature of Auroræ is as yet but little explored. It has been considered necessary to abandon the former view — that they are discharges of the electricity collected at the poles — because it has been hitherto found impossible to bring the chief lines of the Aurora-spectrum into coincidence with the spectra of the atmospheric gases. Theoretical considerations, based on the great alterations to which the spectrum of the same gas is subject under varying conditions of temperature and density, have very recently led Zöllner to the opinion that probably the spectrum of the *Aurora does not coincide with any known spectrum* of the atmospheric gases, only because it is a spectrum of another form of our atmosphere hitherto incapable of artificial demonstration[19].

The following article will show how far I have succeeded, in conjunction with Dr. Lohse, in supporting this view by exact observations of the Aurora-spectrum itself, as well as by comparison with the spectra of the gases constituting the air.

The star-spectrum apparatus belonging to the 11-inch equatorial of the Bothkamp Observatory was used for these observations. It consists of a set of prisms *à vision directe*, five prisms with refracting angle 90°, slit, collimator, and observing telescope. The lowest eyepiece (magnifying four times) of this telescope was employed. The telescope is capable of being moved in such a way, by the aid of a micrometer-screw, that different portions of the spectrum can be brought into the centre of the field of vision. As fractions of the rotation of this screw are marked, the distances of the spectral lines can be readily found.

Repeated measurements of 100 lines of the solar spectrum have enabled me, upon the basis of Ångström's Atlas ('Spectre normal de Soleil'), to express the indications of the screw directly in wave-lengths.

In place of the cross wires originally introduced into the focus of the observing telescope, I have inserted a tiny polished steel cone, the very fine point of which reaches to the centre of the field of vision. The axis of this cone stands perpendicu-

[18] Communicated by the author to the Royal Saxon Academy of Science, 1871.
[19] Reports of the Royal Saxon Academy of Science, Oct. 31, 1871.

lar to the length of the spectrum, therefore parallel with the spectral lines, and the setting of the point of the cone on the latter is accomplished with great sharpness. If the spectrum is very faint, or consists only of bright lines, the cone is lighted by a small lamp. For this purpose, opposite to the point of the cone, there is an opening in the telescope, through which, regulated by a blind, light can be thrown on the point. As the latter is polished, a fine line of light thus appears, which extends to the centre of the field of vision, and the brilliancy of which can be altered by withdrawing the lamp to a greater distance or lowering the blind, so that even the faintest lines of a spectrum can be brought with facility and certainty into coincidence with this line of light.

The head of the micrometer-screw is divided into 100 parts, and each part, in the neighbourhood of the Fraunhofer line F, answers to about ·00016 wavelength. The probable error of position on one of the well-marked lines in the sun's spectrum amounts to about 0·008 of a turn of the screw with the lowest eyepiece of the telescope. I have subjected the screw itself to a thorough examination with reference to such range, as well as to periodical inequalities in the single worms of the screw, but could discover no error exceeding 0·01 of a turn of the screw. I have to mention, further, that after each observation in the position in which the instrument was used, readings followed on the sodium-lines, or on some of the hydrogen-lines, in order to eliminate errors which might arise in the unavoidable disturbance of any particular part of the spectral apparatus.

1. Observations of the Aurora.

1870, Oct. 25th.—A very bright Aurora. In the brightest parts, besides a very bright line between D and E, several other fainter lines were to be discerned, situated further towards the blue end of the spectrum. They appeared on a dimly-lighted ground, and stretched out over the Fraunhofer lines E and *b* to about midway between *b* and F. Towards the red end the spectrum was terminated by the bright line first mentioned. No measurements could be taken, as the apparatus had not yet undergone the above-mentioned alterations, and even the brightest line of the spectrum did not diffuse sufficient light to be able to perceive the fine cross wires. The red rays of the Aurora were not examined.

1871, Feb. 11th.—Towards ten o'clock appeared in the north-west a very bright light-bow of greenish colour as the edge of a dark segment. Even with a very narrow slit, the line between D and E could be well recognized and measured. The average of six readings gave 7·11 turns, equal to 5572 wave-length. In a small spectroscope of low dispersion which is arranged on Browning's plan, a few more lines placed further towards the blue could be recognized (as in October). Towards the red end of the spectrum no lines were observable. The greatest development of the Aurora was about midnight. Magnificent rays rose to about 60° elevation; they had the same greenish colouring as the bow of light, and the appearance of the spectrum also was exactly the same. I again obtained two sets of measurements: the average of six readings in the first set gave 7·10 turns, 5572 wave-length; in another part of the heavens at the same time 7·10 was the result of four readings.

On Feb. 12, towards eight o'clock, the intensity of the Aurora was already great enough to allow measurements of the brightest line. The average of six readings gave 7·09 turns, or 5576 wave-length. Dr. Lohse took observations later, with the same apparatus, and found from six readings 7·12 turns, or 5569 wave-length.

Yet the appearance of the spectrum in the spectroscope of low dispersion was essentially distinct from that of February 11th. The green continuous spectrum was present; it extended from the bright Aurora-line to the lines *b* of the solar spectrum, and was traversed by some bright lines. Between band *b* and F, was another line standing alone, out beyond F, in the blue part of the spectrum, a clear bright stripe; and just before G a very faint broad band of light was perceived.

Amongst the rays which, later on, shot upwards, and were coloured red at their ends, another very intense red line appeared in the spectrum between C and D, yet placed nearer to C[20].

April 9th.—An exceedingly brilliant Aurora, of which the greater development took place in the early morning hours. Magnificent red sheaths rose up to the zenith. The spectrum was like that observed on February 12th, only much more intense, so that the lines could be seen and measured with the larger spectral apparatus. In the brightest part of the Aurora was the dark segment; the spectrum consisted of five lines in the green, and a somewhat indistinct broad line or band in the blue.

The red rays, on the other hand, allowed us to recognize seven lines, whilst the bright line again appeared in the red part of the spectrum. I could not again perceive the faint stripe observed on February 12th, in the vicinity of line G. The mean measurements of four readings on an average, for each line, gave:—

Turns of screw.	Probable errors.	Wave-length.	Probable errors.	Remarks.	
4·62	·0037	6297	·00014	Very bright stripe.	
7·12	9	5569	2	Brightest line of the spectrum; becomes noticeably fainter at appearance of the red line.	On a faintly lighted ground.
7·92	—	5390	—	Extremely faint line; unreliable observation.	
8·71	21	5233	4	Moderately bright.	
8·95	49	5189	9	This line is very bright when the red line appears at the same time, otherwise equal in brilliancy with the preceding one.	
10·06	20	5004	3	Very bright line.	
12·33	—	4694	—	Broad band of light, somewhat less brilliant in the middle.	
12·59	22	4663	—		

---

[20] This red line was first noticed by Zöllner.

12·88	—	4629	3	Very faint in those parts of the Auro-ra in which the red line appears.

April 14th.—Faint Aurora; only the bright line in the green could be recognized in its spectrum. The mean of two readings gave 7·12 turns, or 5569 wavelength.

I append a table of the wave-lengths of the brightest line, as exactly measured on four evenings:—

$$1871, \quad \text{February} \quad 11 \quad 5573$$
$$'' \quad 12 \quad 5573$$
$$\text{April} \quad 9 \quad 5569$$
$$'' \quad 14 \quad 5569$$

Therefore the average result (if only half-weight is allowed to the last observation, because it only depends upon two readings) gives for the wave-length of the brightest line 5571·3, with a probable error of ·000·92. According to Ångström[21], the wave-length of this line is 5567; according to Winlock[22], on the other hand, 5570.

2. *On the Spectra of some Gases in Geissler's Tubes, as well as on the Spectrum of the Atmospheric Air.*

Numerous experiments have been made in order to find out some admitted connexion between the spectrum of the Aurora and the spectra of the principal gases composing the atmosphere. I limit myself to noticing some of the often-repeated observations in Plücker's tubes, which contained oxygen, hydrogen, and nitrogen, as well as the observations of the spectrum of the air under different conditions. The experiments were made with a small inductive apparatus, in which the length of the spark between platinum points in ordinary air was 15 millims. at the most. As Zöllner (in the pamphlet mentioned) comes to the conclusion, that if the development of the light in the Aurora, according to the analogy of gases brought to glow in rarefied spaces, is of an electric nature, it must belong to very low temperature—in order to bring the gases enclosed in the tubes to glow at the lowest possible temperature, I have always employed such weak currents that the gas was only just steadily alight.

The following observations have been repeated often and at various times. The figures are averages of the indications of the micrometer-screw, so that the uncertainty of the figures will, in the rarest cases, amount to no more than 0·015 turn of the screw, and must be reckoned somewhat more highly only in the case of completely faint misty lines. The spectrum apparatus was that described above, and the slit was nearly the same in every experiment, and so narrow that the sodium-lines could be seen separated. The measurements, for the most part, extend only to the Fraunhofer line G, as I feared lest, through further turning the telescope

[21] Recherches sur le Spectre Solaire, p. 42.
[22] American Journal of Science, lxviii. 123.

by means of the micrometer-screw, too great a pressure might be exercised on the worms of the latter.

I. Oxygen.

a. In the narrow part of the Plücker tube.

Screw.	Wave-length.	Remarks.
3·97	6562	Moderately bright.
5·04	6146	Very bright.
6·98	5603	Very bright, misty towards the violet.
8·19	5332	Faint.
8·95	5189	Moderately bright.
10·97	4870	Moderately bright.
11·02	4863	Faint.
11·26	4829	Bright; misty towards the red end of the spectrum.
13·30	4583	Very faint.
14·05	4506	Moderately bright.
15·55	4372	Moderately bright.

b. In the wide part of the Plücker tube.

Screw.	Wave-length.	Remarks.
6·98	5603	Very faint.
8·95	5189	Very bright.
11·26	4829	Moderately bright.

The lines near 3·97 and 11·02 belong to hydrogen. Probably traces of aqueous vapour were present in the tube, which were decomposed by the galvanic current. These two lines are not to be found in a lower temperature in the broad part of the tube. It is striking that the red nitrogen-line near 5·04 is also missing there. In the narrow part of the tube the lines stand out in the green on a very dimly-lighted ground, whilst in the wider part they appear on a perfectly dark ground.

II. Hydrogen.

a. In the narrow part of the tube.

Screw.	Wave-length.	Remarks.
3·98	6558	Very bright.
6·16	5813	Moderately bright, on both sides very faint lines.

7·01	5596	Moderately bright.	On a dimly lighted ground, which becomes fainter towards the violet.
7·18	5555	Moderately bright.	
7·77	5422	Faint.	
8·95	5189	Moderately bright.	On a faint steadily bright ground.
10·03	5008	Faint.	
10·55	4929	Moderately bright.	
11·04	4861	Very bright.	From 11·5 to 12·9 a bright ground, which towards the violet becomes very bright.
12·86	4632	Moderately bright.	
13·32	4581	Very faint.	On a dull ground.
14·05	4506	Very faint.	
15·90	4342	Very bright.	

b. In the broad part of the tube.

Screw.	Wave-length.	Remarks.
5·30	6063	Faint.
7·00	5598	Bright.
8·96	5187	Very bright.
11·28	4828	"
14·04	4507	Moderately bright.

The lines appeared on a perfectly dark ground.

The tube shows in the narrow part the hydrogen-spectrum of the first order; the lines in the green do not coincide with the lines of the nitrogen, though some lines belonging to nitrogen are found. Here, too, most probably small particles of aqueous vapour have been enclosed in the tube and are decomposed. Very striking is the spectrum in the broad part of the tube; nothing is to be seen of the bright shining lines Hα 3·98, Hβ 11·04, Hγ 15·90; on the other hand, four very bright lines and one quite faint one are in the red end of the spectrum, which appear, in opposition to the spectrum of the narrow part, not on a partially lighted, but on an entirely dark ground. The appearance is very striking if we bring the tube in front of the slit; and so, by degrees, at first the light in the narrow part, then the light at the connecting-point of the narrow and wide parts, and, finally, the light in the latter fall upon the slit. At the connecting-point of the wide ends of the tube the three well-known hydrogen-lines decrease in intensity, the continuous ground of some parts of the spectrum disappears, and a new line appears near 11·28, which has about the same brilliancy as Hβ.

A comparison with the spectrum of oxygen shows the bright lines which are in the spectrum in the wide end of the tube as belonging to that element. The heat evolved by the current appears insufficient to bring the hydrogen to glow, whilst by it the oxygen, which is of a more rarefied character, becomes incandescent. An alteration of the direction of the current has no influence on the appearance.

III. Nitrogen.

a. In the narrow part of the tube.

Screw.	Wave-length.	Remarks.	
3·84	6620	Several faint, broad, close lines, increasing in brilliancy as they approach the violet end.	
4·85	6213		
5·30	6063	Broad bright lines, so close together that the intervening spaces appear like fine dark lines. This part of the spectrum is very bright, but not uniform, being brighter towards the violet end.	
5·51	6000		
5·69	5948		
5·87	5896		
6·04	5846		
6·20	5802		
6·43	5741		
6·96	5607	Group of faint but at the same time very broad lines. The last is the brightest.	
7·13	5567		
7·28	5532		
7·55	5470	The dark intervening spaces are somewhat broader, the bright lines somewhat more intense than in the preceding group, and all of almost equal brilliancy.	
7·74	5428		
7·92	5389		
8·09	5353		
8·32	5306	Very faint fine lines.	
8·50	5272		
8·69	5237		
9·01	5178	Very bright broad misty line.	
9·67	5066	Very bright line.	The bright lines are sharply defined towards the red end of the spectrum, fading away towards the other end of the spectrum.
10·25	4975	Very bright line.	
10·66	4913	Very bright line.	
11·03	4862	Very faint line.	
11·41	4811	Bright line.	
12·11	4721	Bright line.	
12·57	4666	Faint line.	
12·57	4644	Bright, broad, misty line.	
13·42	4570	Very bright line.	
14·24	4487	Very bright lline.	
15·02	4417	Bright line.	

15·66	4363	Bright lines.	Bright lines sharply de-fined towards red end, indistinct towards other end of spectrum.
15·72	4357	Bright lines.	
15·87	4345	Bright line.	
16·72	4273	Bright line.	

Here follow several lines.

b. In the wide part of the tube.

Screw.	Wave-length.	Remarks.
6·20	5802	Faint, indistinct, broad line.
7·72	5433	Dull stripe.
8·20	5330	Faint line.
8·94	5191	Very faint line.
9·03	5175	Broad band of light.
9·90	5029	Dull band of light.
10·68	4911	Moderately bright line.
11·42	4809	Faint line.
12·59	4663	Bright line.
13·43	4569	Bright line.
14·07	4504	Moderately bright line.
14·25	4486	Very bright line.
15·85	4347	Very bright line.
16·76	4273	Moderately bright line.

c. At the aura of the negative pole.

Screw.	Wave-length.	Remarks.
5·18	6100	Broad, moderately bright stripe, indistinct at the edges.
5·70	5945	
7·60	5159	Broad, moderately bright stripe.
8·41	5289	
8·76	5224	Very bright line, somewhat indistinct towards the violet.
9·19	5147	Faint line.
10·00	5004	Bright line, indistinct towards the red.
10·67	4912	Somewhat fainter than the last, indistinct towards the red.
11·43	4808	Very faint line.

12·25	4704	Very intense, broad, indistinct towards the violet.
12·73	4646	Very faint line.
13·43	4569	Moderately bright, indistinct towards the violet.
14·25	4486	Like the last.
15·03	4417	Quite a faint line.
15·86	4346	Moderately bright line.
16·76	4275	Very bright line.

Here follow several other lines.

The observations in the different parts of the tube show plainly the dependence of the spectrum on the temperature. The aura of the negative pole gives the line near 10·07 so characteristic of the air-spectrum. This is the same line which is met with in the spectra of most of the nebulæ. The very striking groups of lines in the red and yellow in the spectrum of the narrow part of the tube disappear entirely in the wide part. If we compare the spectra with those above quoted, of oxygen and hydrogen, we find line Hβ very faint in the spectrum of the narrow part of the tube near 11·03; on the other hand, oxygen-lines appear in the broad part near 8·20, 8·94, and 14·07. Thence I would conjecture that the tube was not filled with pure nitrogen, the appearance of which is precise, but with dry rarefied air, since Wüllner's researches have proved that dry air yields the same spectrum as nitrogen gas. Perhaps the air in the tube examined by me had not been thoroughly dried, and thus the appearance of some lines of the elements before named is to be explained.

I must further mention that the electrodes of the tubes consisted of aluminium; yet a comparison of the spectra observed and the aluminium spectrum has shown no connexion between them.

IV. Atmospheric Air.

Screw.		Wave-length.	Remarks.
	5·88	5892	Very bright double line (Na).
	6·67	5680	Very bright line.
	7·20	5550	Faint line.
	9·00	5180	Very bright line.
	9·79	5047	Fine faint line.
	10·03	5008	Very bright double line.
	10·07	5002	
	11·43	4803	Faint confused line.

	12·69	4651	
	12·84	4633	Faint line not sharply defined.
	13·04	4612	
From	14·61	4453	Confused band of light, which ends with a
to	15·88	4444	broad washy line.

Here follow several other lines.

Rarefied air saturated with aqueous vapour.

	Screw.	Wave-length.	Remarks.	
	3·97	6562	Moderately bright line.	
	(5·88)	5892	Bright double line (Na).	
	(6·25)	5789	Bright line (H).	
From	7·03	5591	Broad dull band of light; near 7·03 a some-	
to	7·55	5470	what brighter line.	
	7·59	5461	Bright line (H).	
	8·72	5231	Dull stripe.	
	8·96	5187	Broad misty stripe.	On a dull steady ground.
	10·07	5002	Faint line.	
	11·05	4859	Very bright line.	
	12·21	4709	Moderately bright line.	On dimly lighted ground, becoming fainter towards the violet.
	12·75	4644	Line fainter than the preceding.	
	13·28	5585	Very faint line (H).	
	(15·71)	4358	Very bright line (H).	
	15·90	4341	" "	

Here follow several more lines.

In the first observations, the electric spark, about 1 centim. in length, was allowed to pass between platinum points in ordinary air.

The sodium-line near 5·88 appeared continually. The bright double line at 10·03 and 10·07, with a weaker current or longer spark, was no longer to be recognized as a double line, but appeared as a broad somewhat confused line, of which the brightest part was near 10·05. No lines belonging to the platinum spectrum appeared. Ordinary rarefied air, under a pressure of 25 to 30 millims., and which was enclosed by mercury in a tube 8 millims. wide, showed exactly the same lines as Plücker's nitrogen-tube (*b*), except that some lines belonging to the spectrum of mercury also appeared.

This observation may be regarded as a confirmation of the conjecture above expressed as to the condition of Plücker's tube III. (nitrogen). In the observations described under *b*, the air saturated with aqueous vapour was under a pressure of 22 millims. Besides the sodium-lines, lines of the mercury-spectrum appeared at 6·25, 7·59, and 15·71. The spectrum of rarefied air under similar pressure was found to accord completely with the spectrum of the light in the broad part of Plücker's tube.

III. (Nitrogen *b*.)—A comparison of the spectrum of rarefied air saturated with aqueous vapour with the former shows the striking alterations in the spectrum which are brought about by the presence of the aqueous vapour.

3. Comparison of the Aurora-Spectrum with the Spectra of Atmospheric Gases and of Inorganic Substances.

In the next place, I turn to the comparison of the observed spectra of different gases and of the air with the spectrum of the Aurora. The first band of light in the red part of the Aurora-spectrum most probably coincides with the first system of lines in the spectrum of nitrogen (*a*). Probably only the bright part of this group of lines is perceptible, on account of the extreme faintness of the Aurora; and as in nitrogen the increase of the brilliancy of the spectrum takes place towards the violet end, the absence of the intermediate spectrum towards this direction would find its explanation. The most intense line of the Aurora-spectrum at 7·12 is to be also found in the spectrum of nitrogen (*a*)—as a very faint line, however. That this line appears in the Aurora by itself, and with intensity relatively great, need not appear strange, considering the great alteration of the gas-spectra under different conditions of pressure and temperature. The third line of the Aurora-spectrum, very vaguely defined on account of its great faintness, coincides in the same way with a nitrogen-line.

The line at 8·71 is met with in the nitrogen-spectrum (*c*), as well as in the air-spectrum (*b*). The third line of the oxygen-spectrum at 8·95, which *seems to appear under very different conditions*, is found again, as the fifth line in the spectrum of the Aurora. *Moreover, the sixth line in the Aurora at 10·06 coincides very exactly* with the known nitrogen-line appearing in the spectra of some of the nebulæ. Lastly, as to the broad band of light in the Aurora-spectrum from 12·33 to 12·88, several lines are found in this place in the spectrum of nitrogen as well as the air-spectrum (*a, b*); so that here, too, a coincidence between the spectra may be regarded as probable.

The observations show with some certainty that at least one line at 10·06 of the Aurora-spectrum coincides with the maximum brilliancy of the air-spectrum, and that the other lines appear with great probability in the spectra of atmospheric gases.

In the very great difference of the gas-spectra under varying conditions of pressure and temperature, it would indeed be difficult to succeed in producing artificially a spectrum which should resemble that of the Aurora in all parts. Moreover, it must be admitted, under the hypothesis that the Auroræ are electric discharges in rarefied air-strata, that these strata, qualified for the transmitting of electricity, will have a very considerable thickness.

In this case the conditions of pressure on these air-strata are themselves so different that, within certain limits, each will yield its own peculiar spectrum; but we shall see the sum of collective spectra, so to speak, spread out behind each other; and therefore the impossibility of attaining a perfect agreement between the Aurora-spectrum and the artificially exhibited spectra of mixed gases is evident.

A comparison of the Aurora-spectrum with the spectra of inorganic substances may be easily worked out by the help of the above-quoted wave-lengths of the single lines of the former, with due regard to probable errors, and with the aid of Ångström's Atlas of the Solar Spectrum. Here the perfect harmony of the brightest Aurora-line (which was fixed with an exactitude of about one seventh of the separation of the sodium-lines) with the lines of the iron-spectrum is especially striking. The wave-lengths in the above-cited observations of the bright Aurora-line vary between 556·9 and 557·3, whilst, according to Ångström, two lines of the iron-spectrum are situated at 556·85 and 557·17.

Iron-lines corresponding to the other Aurora-lines, within certain limits of accuracy, are also to be found, as will be seen from the following comparison:—

Aurora-lines.		Lines of the iron-spectrum.	Remarks.
	629·7	630·08	Moderately bright.
		629·85	
	539·0	539·60	Mostly very faint.
		539·92	
		539·05	
		538·85	
	523·3	523·43	Very faint.
		523·21	Moderately bright.
		522·90	Very faint.
	518·9	519·79	Very faint.
		519·40	Very faint.
		519·16	Moderately bright.
		519·06	Moderately bright.
		518·51	Very faint.
	500·4	500·65	Very faint.
		500·52	
		500·49	
		500·30	
		500·20	
From	469·4		3 stronger and 4 very faint iron-lines.
to	462·9		

Yet this agreement, though remarkable, can only be considered as complete proof of the presence of iron-vapour in the atmosphere when we shall have succeeded in showing by observation analogous modifications of the relative conditions of brilliancy in the iron-spectrum by alterations of temperature and density; and in this way explain the appearance of relatively very faint iron-lines in the Aurora-spectrum, or, on the other hand, the absence of the most intense lines.

It will meanwhile remain far more in accordance with probability to regard the *Aurora-spectrum as a modification of the air-spectrum;* since we are already aware, in the case of gases, of the alteration of the spectra by conditions of temperature and pressure; and an agreement, at any rate, quite as certain between the spectrum in question and the spectra of atmospheric gases has been proved above.

[I am indebted to Miss Annie Ludlam for a translation from the German of the above Memoir. —J. R. C.]

Lector House believes that a society develops through a two-fold approach of continuous learning and adaptation, which is derived from the study of classic literary works spread across the historic timeline of literature records. Therefore, we aim at reviving, repairing and redeveloping all those inaccessible or damaged but historically as well as culturally important literature across subjects so that the future generations may have an opportunity to study and learn from past works to embark upon a journey of creating a better future.

This book is a result of an effort made by Lector House towards making a contribution to the preservation and repair of original ancient works which might hold historical significance to the approach of continuous learning across subjects.

HAPPY READING & LEARNING!

LECTOR HOUSE LLP
E-MAIL: lectorpublishing@gmail.com